WORKING ACTORS

WORKING ACTORS

The Craft of Television, Film, and Stage Performance

RICHARD A. BLUM

· · · · · ·

With special thanks to Laurence Frank

Focal Press

Boston · London

Focal Press is an imprint of Butterworth Publishers.

Library of Congress Cataloging-in-Publication Data

Blum, Richard A.
  Working actors : the craft of television, film, and stage performance / Richard A. Blum, with special thanks to Laurence Frank.
    p. cm.
    Bibliography: p.
    Includes index.
    ISBN 0–240–80004–4
    1. Motion picture acting. 2. Acting for television.
3. Acting. 4. Actors—United States—Interviews.
I. Frank, Laurence. II. Title.
PN1995.9.A26B58   1989
792'.028—dc19   88-28225

British Library Cataloguing in Publication Data

Blum, Richard A.
  Working actors : the craft of television,
  film and stage performance.
  1. Acting. Techniques
  I. Title
  792'.028
  ISBN 0–240–80004–4

Photo Credits: Publicity stills, unless otherwise indicated, are used courtesy of the respective actors.

Butterworth Publishers
80 Montvale Avenue
Stoneham, MA 02180

10 9 8 7 6 5 4 3 2 1

Printed in the United States of America

For Jason, Jennifer,

and all who appreciate the

craft of acting

You must be prepared to sacrifice in order to succeed. You must set your goals high and go for them with the pugnacity of a terrier.

—Laurence Olivier, *On Acting,* 362.

The main difference between the art of the actor and all other arts is that every other artist may create whenever he is in the mood of inspiration. But the artist of the stage must be the master of his own inspiration . . . That is the chief secret of our art.

—Constantin Stanislavski, *My Life in Art,* 571.

Contents

Acknowledgments

Writing a book is supposed to be a lonely business. But in this work, I've had the pleasure of working with longtime friends and associates. Larry Frank served as my West Coast interviewer, and I appreciate his time, talent, and prowess. Without him, this book would not be possible. I must also thank Basia and Alexander for allowing him time off to get involved in this venture.

Frank Tavares helped transcribe material, and he had many astute suggestions. Without his adroitness at the computer keyboard, this book might still be in first draft stages.

At Focal Press, my thanks go to Karen Speerstra, Editor, for her professional support and sound advice, and to Phil Sutherland for his constructive editorial feedback and good humor. It's always a positive sign when publishers remain friendly.

Friends and family have been particularly supportive. Thanks to Eve for all the years of generous advice and support, Jason and Jennifer for understanding the demands of writing and still being so caring, Steve for enduring long distance calls, Al for the good memories.

I must thank Norman Tamarkin, whose guidance and advice has always been reliable, reassuring, and on target. He's an extraordinary motivator. My thanks to *Primetime* co-author, Richard Lindheim, for patiently waiting in the wings. Now we'll tackle our next project for Focal Press, a book on creative producing for television.

At the University of Maryland, I wish to express my appreciation for the Arts & Humanities Faculty Research Award, which helped support travel and research activities.

Finally, I must express my appreciation to all the actors who agreed to be interviewed for this volume. Without them, there would be no book. They were personable, intelligent, and informative in their views, insightful in their comments. I tried to maintain the integrity of those independent views in the editing process, and I take full responsibility if there are any misrepresentations. I also appreciate their permission to use the photographs appearing in this book.

WORKING ACTORS

Chapter **1** ·

Introduction

Their names may or may not be familiar, but their faces and characters are instantly recognized by millions of television viewers, motion picture audiences, and theatre goers. They are part of a distinctive class of actors, those who work regularly in their craft.

Working Actors is a forum for successful actors to share their ideas and techniques with readers who wish to explore the intricacies of the craft of performance. The book comprises diverse interviews with television and film actors, many of whom also do stage work. It's our intention to provide a resource for examining a wide spectrum of views about acting styles, techniques, and theories.

This book offers insight into the craft of performance, by those who know it best—actors who have appeared regularly on television, film, and stage. We are not talking about "superstars" who rise meteorically and disappear into the rare stratosphere of stardom. Rather, we are talkng about people who regularly earn their living from working in the craft.

The descriptive term *working actor* is not an idle one. Each year, the acting unions and guilds issue reports of employment statistics, and the results are sobering. The statistics point to perennial employment problems in The Screen Actor's Guild, American Federation of TV & Radio Artists, and Actor's Equity Association.[1]

. . . .

[1] For latest employment statistics, cf. Annual Reports of American Federation of Television & Radio Artists, 6922 Hollywood Blvd., Hollywood, CA 90028; Screen Actor's Guild, 7065 Hollywood Blvd., Hollywood, CA 90028; Actor's Equity Association, 165 W. 46th St., New York, NY 10036.

For that reason, being a working actor is most significant. Despite enormous obstacles, some actors do seem to move with comparative ease from job to job, series to series, movie to movie. This is not to suggest they are immune from job insecurity. That element is built into the very nature of the profession. But there is a core of actors who can be assured of continually working in the craft. How did they accomplish that? We've interviewed some to find out.

There is no direct path toward success for professional actors. Each must be prepared to meet a variety of artistic and pragmatic challenges. Before we look at those acting challenges, we need to understand some basic principles of acting.

Basic Principles of Acting

Although individual approaches and styles differ, actors rely on similar principles to accomplish the demands of the craft. They must be in touch with several aspects of their inner and external resources. These are the actors' tools. They must be finely tuned.

The physical and creative resources of acting are not separate and distinct. On the contrary, they overlap and merge in practice. We might view these principles as holistic in approach. Actors can tap into one area, stimulate others in the process, dismantle others not relevant. The tools are that closely interrelated. These are some of the basic building blocks for actors.

Imagination and Spontaneity

Actors deal with their imaginative processes all the time. They challenge themselves to imagine circumstances that are not real. They ask themselves "What if" this were the case, and they move on from there. They may ask what it's like to be an animal, a character, an inanimate object. The idea is to inspire imagination and creativity. They can draw on their creative resources to create a new character or situation.

Improvisation is rooted on that principle of imagination. It calls on the actor's ability to create structure where there is none. It's a principle resource for sharpening imagination and building up spontaneous values in a performance. With improvisations, no two performances are alike. And that's precisely what actors want in the creation of their roles—a unique performance. Using imaginative skills, they strive for the illusion of immediacy. That sense of spontaneity and freshness is a key objective. Laurence Olivier refers to it as "the olive in the cocktail" of his performance.[2]

Analytic Abilities

These are the intellectual resources used for script analysis. An actor reads a script and must be able to determine the writer's intention for his or her character. What is the character's objective and motivation? How does that character fit into the framework of the entire story? How does the character feel about others in the scene? How do others feel about that character? Those are some of the preliminary questions undertaken in a character analysis.

Analytic skills are also used for investigative research on characters. Doing background research on the historical or biographical nature of a character is important to broaden understanding of the events in the script. If the project is set in a different place or time, it's critical to understand the historical and social context of that character.

Analytic skills are also brought to bear in the "cold reading" situation. In a cold reading, actors audition without time to read the entire script. They're given basic information about the character and situation, but they then must be able to make instant choices concerning a character's attitude, line delivery, interrelationships, and behavior.

Strong analytic abilities are also needed for dealing with out-of-sequence shooting in film. It's important to assess and recall information accurately.

Emotional Discipline

Actors must deal with their own emotions and the inner life of characters being portrayed. They need to be able to draw on their own inner resources at will and to maintain discipline over the emotional aspects of their craft.

Some people draw from their own bank of personal experiences to create appropriate feelings. That approach is called "emotional memory" or "affective memory." They can also substitute their own emotions for those of the characters portrayed. That's aptly called "substitution."

In film, close-up acting requires the ability to listen and react quietly,

• • • • • • •

[2] Laurence Olivier, *On Acting* (New York: Simon and Schuster, 1986), 317.

subtly. People might draw on inner resources to convey that sense of realism to the audience.

For those and other reasons, control and discipline of emotional energies is essential.

Physical Discipline

An actor must be able to create the full spectrum of a character, with external behavior, physical action, and vocal work. We are talking about building a complete physical being, from the way the character walks to the way he or she moves, sits, and sounds.

Often actors use "physicalization" to find out more about their characters. External traits—walk, twitch, dialect—provide clues to who that character is. Physicalization may also provide a route to discover more about the inner life of the character, the way the character thinks and feels.

Acting requires adherence to stage blocking, which is particularly important in television and film. If actors miss their mark, they'll be out of the light, or out of the camera operator's perspective, and the scene must be reshot. Physical adeptness is appreciated by director, cast, and crew.

Actors need to stay in physical shape for the demands of the job. Their training often involves instruction in dance, stage movement, martial arts, voice, and dialects. Classical performers, preparing for Shakespeare, undergo extensive training in stage action, fencing, and vocal techniques.

On the opposite side of this physical pole is the ability to relax, to be in charge of one's entire physical persona. There is a hybrid ability required to relax and to concentrate on the execution of the part.

Psychological Discipline

This principle is admittedly broad. It involves the integration of self with all the other aspects of the craft. It implies the actor is capable of dealing with the craft as a whole and is in control of all emotional, physical, and intellectual resources.

Psychological preparedness puts the actor in control of the unusual demands of this profession. Among them: success and rejection, uncertain job opportunities, long periods of inactivity on a set, intense competition, selling oneself in auditions, and job limitations imposed by typecasting.

No job description is likely to include the principle of psychological discipline, but it is an important resource for maintaining control over the vicissitudes of the craft and the complex demands of acting styles.

Overview of Acting Styles

Acting in television and film requires the ability to make audiences suspend their disbelief and accept the "illusion of reality" in a performance. That's the essence of realistic acting styles.

Not *all* acting styles are realistic. Some, like those of early theatre or silent film, required exaggeration of emotions and physical movements. Some contemporary theatre styles require extraordinary physical ability to create masks out of muscles, or to create anti-naturalistic postures and gestures.

However, television, film, and traditional stage demand a high degree of realism. Characters must look and sound as if they are true to life, reacting in every moment of the scene. Otherwise, audiences simply won't believe it.

Within the realistic acting mode, competing schools spring up like warring factions. Some say actors must begin with their own feelings, others say external work provides that behavioral reality, still others say imagination and analytic abilities are the key.

In the quest for realistic acting styles, America reached back to European theories and controversies dating back to Denis Diderot's "The Paradox of Acting" (1830) and François Talma's "Reflections on Acting" (1877). Diderot believed that actors must portray emotions without actually feeling them. In contrast, Talma felt actors must feel the emotions to portray them honestly.

The basic conflict between physical and inner techniques permeated the history of twentieth century acting styles. At the turn of the century, Constantin Stanislavski, Artistic Director of the Moscow Art Theatre, experimented with techniques to give actors more control over emotions, imagination, and spontaneity. His experiments used both emotional and physical aspects of training.

When his system came to America, it was greeted enthusiastically, but it became the source of great controversy and misunderstanding. Stanislavski experimented with evolving techniques over the years. His early system emphasized inner techniques to build emotional realities. His later theories emphasized the importance of physical actions. Stanislavski meant the actor to be disciplined in both inner and external techniques.[3]

In America, his early techniques were taught by a former student, Richard Boleslavski, who founded the American Laboratory Theatre (1924).[4] His work had great impact on American actors, including Lee Strasberg, who singled out affective memory aspects of the system and used it as the core to understanding Stanislavski.

The Group Theatre was formed in the 1930s by Lee Strasberg, Harold Clurman, and Cheryl Crawford to train American actors in the Stanislavski system. Under Strasberg's direction, the system exclusively dealt with emotional techniques.[5]

During that period, Stella Adler went to Paris to study with Stanislavski. When she returned, she presented the "corrected" view of the system, emphasizing physical actions, circumstances in the play, and the actor's imagination.[6] Most importantly, she stressed working from the script, rather than delving into intimate aspects of actor's lives. The Group members were most receptive to that new approach; Strasberg was not.

To carry on the Group Theatre tradition, Elia Kazan, Robert Lewis, and Cheryl Crawford founded the Actor's Studio (1947). Despite the concern of Robert Lewis and others, Strasberg eventually became artistic director. Under his guidance, the studio exclusively stressed emotional and psychological principles. That style—probing into the inner life and unconscious of the actor—became known as the Method.[7]

It's important to remember that the Method was popularized at a

•   •   •   •   •   •   •

[3] Stanislavski's books detail his ideas in depth. See Bibliography for listings.

[4] Richard Boleslavski's work, *Acting: The First Six Lessons* (New York: Theatre Arts Books, 1949), encompassed a ten-year span of writing about Stanislavski's ideas.

[5] The history of the Group Theatre and Strasberg's work is detailed in Harold Clurman, *The Fervent Years: The Story of the Group Theatre and the Thirties* (New York: Hill and Wang, 1945).

[6] Stella Adler discusses her work with Stanislavski in, "The Reality of Doing," *Tulane Drama Review,* 9, No. 2 (Fall, 1964), 139. For Stanislavski's recollection, see David Garfield, *A Player's Place: The Story of the Actors Studio* (New York: Macmillan Publishing, 1980), 34.

[7] For Strasberg's personal recollections of the Method and the Actors Studio, see Lee Strasberg, *A Dream of Passion: The Development of the Method* (New York: Little, Brown and Company, 1987).

time when Americans were obsessed with psychoanalysis and dramatic realism. It was the 1950s and 1960s. Analysis was intellectually avante garde. The stage was rebelling against traditions, but television and film demanded realism. Tense close-ups told us a great deal about the inner thoughts of our cinematic rebels and heroes. The Actor's Studio and the Method fit naturally into that context.

Dramatic television was in its adolescence, and cinematic achievements were starkly realistic. Elia Kazan, who founded the Actor's Studio, directed Marlon Brando in *A Streetcar Named Desire* and *On the Waterfront*.[8] Studio members Marlon Brando, James Dean, and Marilyn Monroe epitomized the mystique of cliquishness and indulgence associated with Method actors.

Despite the controversy, The Actor's Studio opened branches on both coasts, retaining its hold on Method acting techniques throughout the 1970s and 1980s. In contemporary terms, there is no denying the fact that many of its practitioners have made a major contribution to the field.[9]

However, the same can be said of many actors trained in related techniques. Major stars and aspiring actors have studied under the tutelage of well-known instructors who have refined their own approaches to acting realism. Among the leading teachers: Stella Adler, Sanford Meisner, Robert Lewis, Michael Chekhov, Sonia Moore, Uta Hagen, Herbert Berghoff, Gene Frankel, Curt Conway, and Milton Katselas.

Obviously, there are many acting techniques. Some concentrate on script analysis, others on imagination, others on physical action. Repertory companies offer actors the opportunity to explore different character parts. Classical training, with its emphasis on physical and vocal expertise, prepares actors for the challenge of period plays. Cabaret training affords the opportunity to work one-on-one with an audience.

In New York, where theatre and videotape predominate, there are countless workshops in stage performance, audition techniques, char-

• • • • • • •

[8] Elia Kazan discusses his work with Actor's Studio members, and with Lee Strasberg. He eventually broke from Strasberg because of his increasingly "egocentric interpretation" of Stanislavski, and the fact that Studio actors were being "trapped in introspection." See *Elia Kazan: A Life* (New York: Knopf, 1988).

[9] For a list of members who have won awards in television and film, see Richard A. Blum, *American Film Acting: The Stanislavski Heritage* (Ann Arbor: U.M.I. Research Press, 1984), 79–90. For a list of life members of the Studio, see David Garfield, *A Player's Place: The Story of the Actor's Studio*, 277–280.

acter analysis, and taped scenes for daytime drama and commercials. In Los Angeles, where television and film acting is concentrated, a variety of workshops are offered in the pragmatics of cold readings, auditioning, scene preparation, and acting for the camera.

Actors have to be adequately prepared to deal with the myriad of artistic and practical challenges faced in the craft. No matter what the acting style, the end result must be the same—a credible performance for the audience.

Special Aspects of Television and Film Performance

In television and film, there are specialized demands placed upon the actor, including the ability to deal with film directors and crew, camera coverage, out-of-sequence shooting, and limited rehearsal time.

The psychological discipline referred to earlier is especially evident in the unique demands of television and film acting. These media require spontaneity in repetitive takes of scenes and the ability to build emotional "matches" from take to take. Creating that sense of spontaneity in each performance is critical for the performance to work credibly in the final, edited version.

Another psychological dimension is the ability to cope with the insecurities of the profession. Actors must be resilient in the face of long periods of unemployment when they are unsure of what job is next. Actors in network series face hiatus and eventual cancellation of their series.

What techniques are used by those working successfully in the craft to deal with these situations? That's what we've set out to uncover for you.

Topics Covered in Acting Interviews

We tried to cover the topics most often asked by new actors and professionals at work in the craft. The questions were designed to elicit a common framework for discussion. Some topics were naturally more relevant to certain actors than others. Their responses reflect those individual differences.

Background and Training

At the beginning of the interview, each actor was asked to discuss his or her own background. Sources of training and influences are naturally important for understanding later career experiences. Consequently, we discussed training, education, and early career influences.

Analyzing the Script and Part

Actors were asked how they approached new scripts and characters. What happens when they first bring the script home? If they've been playing a part for a long run, how do they keep characters fresh?

Researching and Building Characters

Researching and investigating character is a vital tool for character preparation. We asked about specific preparation required before coming to the first rehearsal. In cases of different physical character types, we discussed techniques for physicalizing roles.

Working with Method Actors

Since the Method and its variants are still rooted in the system, we asked for reactions to working with Method-trained actors. We also discussed the Method as a tool for preparation.

Approaching Drama and Comedy

Some actors have specialties in drama, comedy, or both. We explored individual approaches to situation comedy, dramatic film, and stage work.

Rehearsals

We asked about problems in the rehearsal process. In television and film, there is considerably less time for rehearsals than in stage. We also explored the use of improvisations in rehearsals and on the set.

Working with Directors

In television and film, there is relatively limited time to work on the problems of a part. Given that reality, we discussed the intricacies of actor-director relationships.

Comparing Acting Environments

Many television and film actors also work on stage. Theatre is perceived as a place to sharpen timing, with immediate feedback from live audiences. We asked for specific comparisons of acting environments in television, motion pictures, and theatre.

Production Pragmatics in Television and Film

Acting for the camera has some unique demands, from close-ups to repetitive takes of scenes. ("Takes" assure the director there will be sufficient coverage of that scene for editing purposes.) Some actors preferred many takes, others as few as possible. Another production demand: out-of-sequence shooting. That situation requires actors to recall precisely where the character is in the script, and to build consistency of character. We talked about these and other problems associated with the actual production process.

Selecting the Next Part

Criteria for selecting new projects were discussed, as well as the problem of "typecasting," which restricts offers of new parts. Some actors wanted to play similar parts, others wanted to stretch into new roles.

Casting

Do well-known actors have to go through the audition process? How do they deal with casting? We asked them for specific reactions to casting and cold reading situations.

Insecurities of the Profession

All actors face the prospect of periodic unemployment. Series stars deal with hiatus between production seasons. Other actors deal with extended periods of inactivity "between jobs." We asked each to discuss means of coping with that reality.

Advice to New Actors

We asked about advice they would offer to new actors concerning training, breaking in, and becoming a working actor.

Personal Assessment

We asked several actors to assess their own individual attributes. What personal characteristics did they feel contributed to their success?

Working Actors: A Resource of Experience

The actors interviewed have a collective experience that represents hundreds of years of knowledge. Their credits in television, film, and stage run into the thousands. Among the areas represented in this single volume—drama, comedy, episodic series, miniseries, television films, motion pictures, soap opera, musical theatre, classical theatre, dramatic theatre, animation, ventriloquism.

Their perspectives provide a vast body of knowledge about the craft. They've taken disparate routes toward success. They were trained in vaudeville, classic theatre, the Method, in college, in workshops, on the job.

Although diversity is the norm, several people in this book had leading roles in the early ABC comedy series, "Soap." They were selected without regard to that fact, but coincidence brings those series alumni back together in this volume. The other actors interviewed help broaden the cross-sectional representation.

The working actors interviewed for this book are in a unique position to offer you critical perspectives of the craft. What training did they find most useful? How have they managed the creative and pragmatic demands of the craft? What advice do they have for new actors?

In these edited interviews, we've tried to retain the personal voices of each actor. In the process, fascinating experiences are uncovered. Each interview represents unique experiences, opinions, and attitudes. You may find similar strains, or contradictory messages. The craft of acting is that complex.

One fact remains constant. Each person in this book was committed to sharing professional experiences and insights with you.

Chapter **2** ·

Dennis Patrick

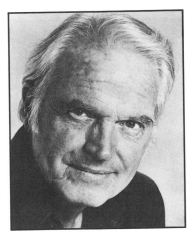

Dennis Patrick talks about the techniques he learned from legendary comics Frank Fay, Milton Berle, and Bert Lahr, and he discusses the importance of the "illusion of the first time." He offers two contrasting examples of how he researches roles and explains why "bastardized Stanislavski" affected the craft. He stresses the importance of listening in film acting and shows how Elizabeth Taylor and Richard Burton had problems in that area. He also offers extensive advice to actors. Dennis Patrick was interviewed in his home.

. . . .

Career Profile Dennis Patrick, actor and director, has performed on stage since the 1940s. He's toured with all major stock companies, played every major city in the United States, and won the Theatre World Award for his work on Broadway. He's portrayed the lives of historical figures including Father Flanagan, Sir Thomas More, Martin Luther, Calvin, Erasmus, Marco Polo, and Thomas Edison. He was a pioneer in the days of live television, and he guest-starred in over two thousand television productions.

A teenager during the Depression, he worked a number of odd jobs, then landed a role in *Seven Mirrors* at the only extant Off Broadway theatre, The Blackfriars. Starting in summer stock in 1947, Mr. Patrick appeared in *Dear Ruth, George Washington Slept Here, Charley's Aunt,* and on Broadway in *Harvey.* Among his other stage performances, *Red Roses for Me, The Silver Tassel, Cock-A-Doodle Dandy,* and *The Hasty Heart.*

He received the Theatre World Award for his performance in *The Wayward Saint*. He also appeared in theatrical productions of Bernard Shaw's *St. Joan*, Carlo Goldini's *The Liar*, and Cole Porter's *Kiss Me Kate*. Among his later stage credits, *Plaza Suite*, *The Andersonville Trial*, *Marat de Sade*, *The Little Hut*, *Tender Trap*, *Children*, *Children*, and *The Seagull*.

Mr. Patrick is a particularly well-known figure in television. In the early days of the medium, he was cast in the first adaptations of *Macbeth*, *Kidnapped*, and *My Heart's in the Highlands*. He came to Hollywood as an actor-director for NBC-TV, directing over fifty network shows, including the Emmy winning "A Gift of Tears." He's worked regularly in the medium since.

As an actor, he's had running roles in television series including "Bert D'Angelo, Superstar," "The American Girls," and "Eight is Enough." He's played the role of banker Vaughn Leland in the CBS-TV series, "Dallas," and he created the role of Patrick Chapin, the patriarch of a wealthy Virginia family in the primetime soap opera, "Rituals." He's also been seen in a number of television films including the acclaimed "Missiles of October," and "Wet Gold."

Among the motion pictures in which Mr. Patrick has been featured, *C Man*, *The Painting*, *House of Dark Shadows*, *Death Squad*, *Tiger by the Tail*, *Deadly Honeymoon*, *Daddy Goes A-Hunting*, *The Jungle*, *The Secret of Nikola Tesla*, *Joe*, and *Major Dundee*.

He's shared stage, screen, and television with Martin Balsam, Charlton Heston, Walter Matthau, Ann Harding, Joan Fontaine, Martha Raye, Joan Bennett, Jean Stapleton, Rod Steiger, Jack Warden, Carol O'Connor, Maureen Stapleton, Julie Harris, Rock Hudson, and many others.

• • • • • • •

You were trained by some of the greatest comedians and stage performers of vaudeville. Can you tell us about those early experiences?
I learned from people like Frank Fay, Milton Berle, and Bert Lahr. When I was a kid, they took me under their wing. If there was any way of making something funny, *they* knew the way. They taught me all about timing. They'd say, "OK, kid, say the line, walk over here, gimme a double take, go out and shut the door—slam it. You'll get a laugh, you'll get a hand." They were always right.

That's my formal training. In essence, I learned while I was working with the great comics. I was an understudy in the play *Harvey*, and one day the guy who played the cabdriver was out doing a movie; so was his first understudy. I wasn't understudying that role, but I memorized it very quickly. The audience was already in the house, I remember, behind the curtain.

It wasn't a very big role, but it was a very *funny* role. The character had a great line, "After this, lady, he'll be a perfectly normal human being. And you know what bastards they are." Well, I wanted to get to that line. It's a "hand" line. That's where you walk out the door, and the audience explodes. You get a spontaneous burst of applause. So, I did it, and I got the hand. You couldn't miss, if you knew how to deliver it.

The techniques I learned from the greats gave me the tools to project and deliver. There's no sense in being a very sensitive and talented young man if you can't be heard beyond the third row in the theatre, right? So you learn to project that damned voice. You don't have to have a big voice, you just do have to have a technique.

Technique for the great comics had to do with how you moved on stage. You never made a stage cross starting out with your downstage foot. It reminded me of my high school drama teacher. He made us walk back and forth across the stage, by the hour. If I crossed stage left, I had to start with my left foot. Right, right foot. He also made us stand still—for hours—while somebody else spoke. Then, he'd make us deliver speeches to see if we'd creep forward as we spoke. (Laughs.) It sure taught us about focus.

So I was taught the importance of physical movement and focus on stage. All the old pros taught me the same thing. If I lean towards you, the audience will look at you to find out what you're going to say. They don't listen to what I'm going to say at all.

Some of them would kill you if you upstaged them. Bert Lahr would put a knife in your back if you so much as lifted an eyebrow when he was coming on with his funny shtick.

Frank Fay would say, "Arm's length, please. A great star is passing." When I was on stage with him, he'd have us all at arm's length. He was insistent that we don't move while he talked, or nobody would hear what he had to say. Once, he was on stage at the Palace Theatre and realized the laughter wasn't all his. Out of the corner of his eye, he saw comic Burt Wheeler chomping on a sandwich. Frank turned around, bowed to Burt, then turned back to the audience, saying "You know, some people will be cute, till they drop dead." In the very next skit, it was Burt Wheeler's turn, and Frank was put down, hard and funny.

W.C. Fields was once upstaged by Ed Wynn, and the results were hilarious. Fields was doing a bit on stage with a pool cue. And underneath the pool table, Ed Wynn was making all kinds of monkey faces at the audience. Fields realized he wasn't getting all the laughs, and when he caught on, he just leaned over, and knocked out Wynn with some ad lib that the audience loved. He won back the focus.

The comics lived and died by winning—and holding—audience focus.

When you first get a script, how do you begin your analysis of the character?
First, I read the script. I've been asked to play or read for a specific part, so I know what I'm looking for. I'm fortunate because I have a facility for reading. I was taught to read by my mother when I was three, so today I can read very quickly. It's a very strong attribute for an actor. I see the entire page, read whole sentences, not just one word at a time. I can also read the dialogue after mine.

In television and film, you have very short scenes. You don't have to memorize a great deal. There's also little time for rehearsal, and very little direction. So it's up to me to know what I want before I get to the set.

When I *do* start memorizing, I break the script down into the longest speeches, and memorize those first. I learned a long time ago that long speeches are easiest to memorize. I could learn those speeches and say my lines in a way that would be simple to repeat. When I moved onto the sound stage, to "loop" it would be easy. I didn't give myself complicated things to do.

You begin to memorize the longest speeches first. What happens if there are substantial script revisions?
That happens a lot, especially in television. You get a script, then the whole script changes. A guy comes around at nine in the morning, rings your doorbell, and hands you a multicolored rainbow. (Laughs.) Different colors for different revisions. Your whole damn speech has been changed.

They can cut a line, or substitute a line. But, if the other character says something differently, or if the lines are in different juxtapositions, that impinges on script learning.

You've played such interesting and diverse characters over the years. Can you give us some specific examples of how you research your characters?
Well, I have two stories about that. I played Thomas Edison in a film

shot in Yugoslavia, *The Secret of Nikola Tesla*. Orson Welles played J.P. Morgan, and Strother Martin played George Westinghouse. When I got the role, I went to the library to research Tom Edison. I found he was a vulgar, smelly old man with a tremendous ego, and he always smoked a cigar.

There's a scene in the film where Nikola, a young Yugoslavian genius, comes to America to meet me. He's in my home, so are George Westinghouse and his wife. I say, "Well, show him in, show him in," and I'm smoking a cigar. Edison always smoked a cigar. He chewed tobacco. He stank most of the time. So this boy comes in and hands me a letter of introduction. I'm saying, "Yeah, well come around Monday morning, and we'll start you off and see what you can do." And I blow my cigar smoke in his face.

The director was European and was very uneasy with the characterization. He didn't think it was appropriate for a great American scientist to treat the Yugoslavian boy like that. (Laughs.) I said, "Well, I researched it, and that's the way the old son of a bitch was. He was mean, he was cantankerous, he was rude and crude."

The director was insistent that blowing cigar smoke at the boy wouldn't be accepted in Yugoslavia. He was convinced that Europeans would think he—the director—was purposely undermining the image of a great American. It would be perceived as an insult. So I had to take it out.

Another time, at Theatre West, I did a scene from the play *Luther*. Luther starts out in the pulpit trying to explain his inability to reach God. He's a very strange kind of character. He talks about sitting on the pot, and looking down and feeling that the rats are coming up from underneath the sewer to grab his testicles. As the speech goes on, he begins, in a sense, to froth at the mouth. Finally, he falls over backward in a faint. When I did that fall onstage, several actors rushed up to help. I snarled, "Get off, I'm acting."

Well, at the end of this one, some of the actors in Theatre West asked how I researched the part. On stage, I fooled the hell out of them, and they wanted to know what kind of research I did to prepare. I told the God's honest truth. I said, "I read the character description by the author." Curt Conway, who ran the acting workshops, looked at the other actors and repeated this heresy, "His research consisted of reading a description by the author." (Laughs.) Then he closed the book and said, "That's all for tonight."

So, that's it for research. You can go through hell digging for clues, or you can find it immediately in a thumbnail.

When you play roles, you're bringing your physical being into the role. You also bring an understanding of the character's attitudes and intentions?

Yes. I am physically who I am, and I use my exact speech to make the character come to life. Those are my basic tools. Generally, I'm cast in something that the audience will accept me in automatically. I don't have to walk in and create a physically different person. I walk in and the audience says, "Here he comes, he's the bad guy," or, "he's the good guy," or "he's the rich guy." Whatever it is, by my demeanor, by the entrance, they know who I am. So, I'm not really creating a physically different character. I am playing myself.

As for understanding the character, you must always believe that the character is right, no matter what. You've got to think that you are right even when you play a "bad guy." To play a real villain with any conviction, you must see your side of it and no one else's.

If you get good scripts with wonderful things to say and subtle things to do, and if you can put the picture of irony in there, you'll have a good time with your character.

What technique do you employ to build your character and to create a sense of spontaneity in your performances?

I've found out that all acting can be brought down to one phrase—"*The illusion of the first time.*"[1] After all, acting is an interpretive art. Someone else wrote this text, created the character. I look at it for the first time in my life—these words printed on a page. They go into my brain and eventually, out of my mouth, memorized. What miracle occurs that makes me say that line, right from the heart, through my mouth, in response to somebody else's words?

This is how the illusion can be crafted. In a play you rehearse until it's instinctive. One actor says a word, you say yours. You work on that, until you get it right. And in that give and take process, you eventually create an illusion of it happening for the first time. It's as if you were actually saying these words and listening to them for the first time. It conveys a sense of spontaneity.

There are a great many acting schools that teach variations of Stanislavski's system. The most prominent is the Actor's Studio and the Method. What are your thoughts about the Method?

•　•　•　•　•　•

[1] This concept was initially proposed by William Gillette, *The Illusion of the First Time in Acting, Papers on Acting I.* (New York: Dramatic Museum of Columbia University, 1915; rpt. 1958.)

(Deep breath.) The Method. It's entirely misinterpreted in this country. I would advise your readers to read the books that came out of the lectures by Robert Lewis called, *Method or Madness?*[2] He found that the very actors he had taught could not play in the plays he wanted to direct. He couldn't have a guy with a heavy New York accent playing Henry the Fifth.

And Stanislavski. Well, they forget Stanislavski was a scientist. He was connected with the greatest single acting company in the world—The Moscow Art Theatre. His country had great respect for actors, supported a national theatre. His company had one goal, to act and explore the craft of acting. Stanislavski worked with all the great actors of his time. And he observed them like a scientist, writing his observations down. He wrote books based on his observations, vital to actors.[3]

He was dealing with some of the best actors in the Slavic tradition. They stretched as actors, portraying many different roles and characters. I think Americans are more visceral in their acting than their European counterparts. Americans somehow feel more comfortable being themselves.

Now that's not necessarily bad. I'll give you an example of a visceral actor. Paul Muni. He used to say. "I act with my stomach." That's the only way he could describe what he did. He once did a live television drama, and could hardly remember his name. They put a small earplug in his ear and gave him his line, each line, from the control room. You would never know it. He was *brilliant.*

You're a member of the Actor's Studio. What experiences have you had working with Method actors?
I'm a member of the playwright's group of the Actor's Studio, and I have directed four plays at the Actor's Studio. I have *never* used an Actor's Studio actor. It just takes too much time. They constantly talk and discuss the scene, the characters, the motivations, for hours and hours. When they *do* productions at the Studio, they still discuss everything to death.

In the American theater there's no time for that. If I had all day long with nothing else to do, and I wanted to experiment, I would take what I call "bastardized Stanislavski," and work with them. That's what I think the Actor's Studio gave to the American theatre—"bastardized Stanislavski." And it damn near killed us.

• • • • • • •

2 Robert Lewis, *Method or Madness?* (New York: Samuel French, 1958). Also see Robert Lewis, *Advice to the Players* (New York: Harper & Row, 1980).

3 For Stanislavski's books, see Bibliography.

The so-called Method killed off a lot of spontaneity. One of their techniques is to substitute emotions. They say, "Well, think of the death of your father, death of a dog, somebody that meant something to you that will help you cry." And they can all cry their eyes out. How they cry! (Laughs.) Bobby Lewis pleaded with one actor who always cried, "Now get to the point of tears, but for God sake, don't cry!"

When I was with Theatre West, they were all dedicated to the Method. They had all gone to the New School of Social Research in New York and had all been taught by Sandy Meisner a form of the Method. And they were devoted to it.

Curt Conway was teaching there, and even he was finding it painfully slow and dull to watch an actor sitting under a tree being an apple for an hour and a half. It's pretty damned dull if you're interested in getting on with it. So he invited me and Carroll O'Conner to meet the group. We were two theatrically oriented actors—guys who came in and memorized the script and got on the stage and did it.

Oddly enough, we all learned from one another. I remember an audition piece I did the first time I was brought in to meet the group. I did John Donne's poem, "Go and catch a falling star / Get with child a mandrake root / Tell me where all good things are / Or who cleft the devil's foot." I recite those things well, with Shakespearean style.

One actor got up and said, "I loved it, it was great. Why don't you do it one more time like a drunk in a bar telling your friend about this woman you lost." Well that was heresy to me, and yet it intrigued me. So, I said, "Well, I don't think I can do that right now, but I'll bring it in next week." And I did. I did it like a drunk. And I was very proud of it. Very happy. So those kind of experiments *can* help you a lot.

Is your approach to comedy different? How important is comedic timing for you?
Comedy is much harder than drama. In comedy, if I personally have to *be* funny, I wouldn't know how. But, if the script is ripe with comedy, and I have great lines, I can get the laughter. I have to work like hell on it, but I can do it.

Sid Caesar says comedy is like two musical instruments. To get a laugh, you have a tympano and you have a triangle. You can get a laugh by going BUM-BADA-BOOM-BOOM-BOOM—tink. Or by going Tink-tada-tink-tink—BAM! Either way, laughter is always tripped. It's a very specific rhythm.

To me, laughter in the theatre is the ultimate proof of a trained talent on the stage. It's like learning to sing pianissimo when you are an opera star. They go through all their lives trying to make their wonderful voice loud enough to be heard ten thousand rows back. And the final thing they learn is how to sing softly, because it requires ten times the control to get a tiny little note to reach those back rows than it does to shout, to sing, to bombast. Comedy and laughter is that to an actor.

You've worked extensively on stage, film, and television. Do you have any preference?
Every actor loves the theatre. I love it, too, but not the hours anymore. (Laughs.) And today, it seems, there's a great deal of amateurism. There's usually somebody in a production who doesn't know what's going on. You either have to surmount it, overcome it, or quit.

That rarely happens in the technical world of film, tape, or television. In a motion picture studio, especially in Hollywood, you're working with the finest technicians in the world. Those of us who came off the Broadway stage a long time ago and now make our living out here, all know that the geniuses of Hollywood are the technicians.

Technical crews also appreciate "good actors," down to the wranglers of the westerns. If you're a good actor, they'll give you a good horse. (Laughs.) And if that horse has a tendency to throw you, they'll work on it until it's tame.

What kind of problems are there in actor-director relationships on the set?
Many young directors don't know what they want. Or, they don't know when they've captured it, so they'll shoot the scene over and over. They tend to know the technical aspects of the production. They know their film, and they know the angle of the picture they want.

They hire actors like me, because of reputation. They tell us what to do, we do it. And that's why people like my work. I come in, memorize, look around. I find out what the blocking is, and work with it.

Sometimes a director is unprepared, and says, "Get up and walk around, I'll see what you want to do." When that happens, it makes me cringe. I get angry with lazy directors.

Close-ups are an essential part of film and television acting. It requires an ability to "listen" and react. Can you talk about that?
All of us who are in the film business know that reaction tells more to an audience than action. It's fine to have speeches, but if they keep cutting to you listening, you must come alive.

I'll give you an example. In the film *Cleopatra,* Elizabeth Taylor did not know what she was doing in the role. She was willing, and she tried, but she was so in love with Richard Burton at the time that she couldn't see straight. Now, Burton was a *magnificent* actor—but he didn't know the film medium at the time. With his great theatrical voice, he didn't know how to handle his body, did not know how to listen. Every time they cut to him, he wasn't doing anything, so they didn't use it. Or if they *had* to use it, you could see his inattention.

In direct contrast, there was Rex Harrison. *Cleopatra* is filled with shots of Rex Harrison listening. And he steals the whole damn film—listening, without uttering a word. Clever man.

I'll give you another example of the importance of listening. I did a television episode with a woman who has the lead in a series. She knows she's important, because she's got a big trailer, a big salary, and the lead. But, she doesn't listen.

I had several scenes with her in which she had nothing but negatives to say. All during my speech, before she heard why she had to say no, she was shaking her head "no," getting ready to say it. Finally, I convinced the director not to do the scene in two shots, just to do them in single close-ups. In that way I could finish my speech without looking at a girl who was saying, "No, no" through the whole thing just so she could say, "*No!*" at the end of my speech. She couldn't listen.

What about the problem of shooting out of sequence? How do you maintain character consistency from scene to scene?
It's not a problem for me as an actor in a regular drama. That's the director's problem. And since I've directed a lot of them, I can tell you it's a *real* problem for the director.

The secret to motion picture direction is to have one damn good memory. Putting these things together in juxtaposition can be maddening: Did it end on an upnote? Should the next scene be picked up on an upnote? Bring it back down? Did it end on a downnote? Should it be picked up in the same mood? Should there be change? The director has to remember constantly where everything was left in sequence.

For the actor, it's no problem. The hardest thing for the actor, in a sense, is *waiting*. It can be very tiring waiting for the scene to be shot. And when it's time, you've got to come to life immediately.

How do actors deal with that period of waiting, and still keep energy levels high?
The key is energy. Jack Lemmon has tremendous energy. If he's sound asleep in his dressing room and they suddenly knock, "You're on, Mr.

Lemmon,'' out he comes. He's right on target, walking, talking, listening. With great energy.

Lee Marvin used to snap his fingers all the time. He was always jiggling on the set, pacing. He kept himself going physically so he could be hyped up for the scene.

Jimmy Cagney once told me, ''All this stuff about relaxation—if you relax, *they,*'' meaning the audience, ''will relax right along with you. And you'll soon hear a lot of snoring.'' If an actor had the line, ''Good morning,'' Cagney would tell him to feel like he's going to shout, ''Run! The house is on fire!'' The line will roar out. It's got to be energized on the inside, but stay in control. That was Jimmy Cagney, all the time. Energized and in control.

When you come to the set, what is expected of you by the director?
Most of us who are successful in this business—those of us who are called upon to work all the time—come in partially memorized. I break the back of the big speech, then get to the little ones.

When we walk on the set, we get a couple of run throughs. The director blocks it out and says, ''Okay, this is going to be a take.'' We know our lines, but we really don't know it like this (snaps fingers). I use that moment to recapture my lines, and in that second of searching, it gives the illusion of my thinking of it for the first time. And hopefully the director catches that credible moment on film.

Do you find improvisation helpful or not?
I have nothing against improvisations. Jack Albertson and I would do improvisation based on one word. We'd do drama. Comedy. Anything! Does it actually help? It's creative, but it doesn't help you memorize a line. It doesn't help you develop a character. It doesn't help you maintain the discipline of the theatre, or of acting in sequence.

You prefer to be true to the writer's text. How far do you think actors should be allowed to move from the writer's dialogue?
The actor should never be more important than the author. The true creative force in the theatre is the author—to the last syllable. I was brought up on Broadway, where the author's contract was inviolate. You could not deviate from the script by a comma.

Broadway still maintains that tradition, although the Actor's Studio destroyed that respect for the author's words. The author had the legal authority to make you say it as it appeared in the script. If you didn't, you were fired. There was no defense you could take.

I started in the classics, with Shakespeare. So I was taught that you

memorize every "if," "and," and "but." With more contemporary television and motion picture scripts, it's admittedly different. Who's
going to bother you about the inversion of a line like, "Hey, Charlie,
where're you going?" instead of, "Where're you going, Charlie?"

There are still some actors who insist on having dialogue directors.
I always say to the script supervisor, "You can't hurt my feelings. Tell
me every 'if,' 'and,' and 'but.' " I'm happier that way, because that's
the way I memorized it.

*Did you ever have problems with directors about interpretations of
character or camera coverage?*
I've never really had any trouble with them. Sometimes you run into
a director who talks too much. They'll end up talking so much, you
don't know what the hell they were asking you to do in the first place.
(Laughs.) They mean well.

There was a wonderful director who never understood why actors
were so angry with him. He *loved* actors, and he could never understand the frustration he caused by his directing style. He spent a great
deal of time explaining the scene, from beginning to end. He explored
every motivation, everything that should be in your head. He'd ask if
we got it all, and we'd say, "Yep."

That's when he pulled the rug. He'd say. "Okay, we'll start with the
close-up at the end of the scene." (Mock horror.) The actors were ready
to *strangle* him. In the motion picture business, the close-up is everything. The close-up should come last in the shooting sequence. By the
time you get to the close-up, you've done it so many times it's down
cold. You can change a performance in a close-up. You can become
more subtle. You can become more certain.

*How do you handle the anxiety of being between jobs? We're told even
Jimmy Stewart worried if he'd ever work again. How do you deal with
that insecurity?*
It's interesting, Jimmy Stewart's insecurity even spilled over to the
stage. When he played Harvey, he'd say, "Can't you get closer? It feels
like everybody's so far away." He didn't feel secure, so he wanted the
actors close to him. Of course at six feet five, what the hell, he was
going to stick out anyway. Isn't that interesting? It just occurred to me
when you asked that question.

Hank Fonda had that problem, too. Especially worried about periods
of unemployment. That's why he did all those plays. He couldn't stand
to be out of work. Some of the greatest actors in the world are shy

people. Helen Hayes said that if she'd ever had to read for a part, she never would have gotten a job as an actress, because she's so nervous about it.

Recently, we had a guest in our home. She's the dean of drama at a university down south, and a fine actress. We were talking about this very thing. She said, "I never could have done without that weekly paycheck. Therefore, I didn't belong in the theatre. You do." We both made a niche, in different ways.

I'm lucky because I've learned to live with it. Even hiatus would be welcome for me. I've never done a full series, so all I've ever known throughout my life is unemployment. (Laughs.) For two thousand weekends I didn't think I would ever work again. So, the guy who works for twelve weeks or twenty-two weeks, and then has six whole weeks off for hiatus—knowing he's going back to work at the end of it—it must be heaven!

Even though it's not a real problem for me, I've had my blue moments. Barbara, my wife, reminds me that for forty-five years I've made a living, which is a miracle. (Smiles.) I'm probably still good for a few more years.

Have you noticed any difference in the parts offered to you over the years?
I've been thinking about that a great deal. In the old days, television plays were morality plays, and we were all cast as bad guys. When we did "Martin Kane, The Private Eye," Mickey Alpert cast all the old vaudevillians in those parts. And they'd always sidle up to you and ask, "Who are you playing—the shit or his friend?" The big roles went to the bad guys, because the hero was too busy being dashing. They needed well-written, well-crafted villains in each episode. That's the part they'd pay for.

Now I'm an older man. The shows are not about me. They're about young girls with big boobs, or young guys looking sterling and wonderful, who run down the street and breathe hard and hop into cars and over motorcycles, and things like that.

I've come to one conclusion. At my age, those of us who do work are hired because we're the "actors." They appreciate us in Hollywood, and they pay us whatever they can afford. And we are treated with a great deal of respect.

But, this is a town that once made a star out of Rin Tin Tin. (Laughs.) And if the dog is bringing in the money at the box office, he get's the trailer and you get the closet!

What advice would you give to young actors just starting out? Any tips about training, breaking in, surviving?

I love to be asked for advice. There's an instinct in all of us as we get older to give the benefit of our experiences to others. I would say to young people, remember age is wisdom. It represents experience. Avail yourself of it. Watch the experienced people. Listen to what they have to say, and try to put some of it into your own life. You will benefit a great deal.

Do whatever you can do to make opportunities happen. After all, we don't have a true theatrical culture in this country. Our forefathers forgot one very important thing, man does not live by bread alone. We're a working class society that's working. But, they didn't put any money out for the actor.

At the Actor's Studio, Bobby Lewis was asked, "How do you get a job? How do you get started?" He said, "Well, I've learned over the years, there's only one thing I can advise you to do. Be there." He let it lay right there. That's just about all you can say about the opportunities: Be there.

Paul Newman is asked this question a lot. And all he ever says is, "Look, Buddy, it's all luck." There are many talented actors who never had a break. They weren't standing on the right street corner at the right time. I've always loved Paul for that. That's his only answer. It's luck. It is. And also the fact that he also happens to be a very brilliant actor.

There are so many new actors coming now. And they're well trained. They're coming out of universities where they're grounded in the classics. I think that's very important, training in classics. It's important to be trained in everything from the classical theatre and physical techniques to the nuances of comedy.

What do you think it is about you that enabled you to beat the odds and be a working actor for these forty-five years?

I learned early on how to be myself. It never occurred to me to play anyone else. I won the Theatre World Award for playing a character who talked to birds and animals. He was an Irishman I grew up with. I knew him inside out. I knew exactly the way he walked, the way he jumped around, everything else. But I was still playing myself.

The other thing is, economically I survived through versatility. I did commercial films in the days when there were no commercials. They took you out to Detroit and paid you $750 to $1,000 to do a show. I once did a thing for Truscon Steel Sliding Closet Door Units. Say that fast three time, and you've earned your money. (Laughs.) I've learned how to do it.

And voiceovers. Those are the highest paid people in our profession. The million dollar a year boys are not always in front of the screen. They're selling the product, the voiceover in your commercial.

I learned how to be a Master of Ceremonies. I was a big guy with the police, hosted their awards dinners. Those things are important. They promote presence on stage, without a specific script. It gives you a maturity and a sense of control in front of an audience.

In the late 1940s if you didn't have a job on Broadway, and you didn't get a job in summer stock, and you didn't have a job in radio—and very few people, contrary to what popular belief is, made any bucks out of radio—you went to work in a department store. Marlon Brando, after one of his biggest hits on Broadway, was an elevator operator in a hotel. You'd take any job that would allow you to go in for an audition somewhere without losing the job and the bread and butter it gave to you.

Along came television. I was very lucky to be at the forefront of television. Plays were transferred directly from the stage, and it allowed people like me to get the experience of being a "star." If you carried the lead in one of those things for an hour, it seasoned you. I don't know how to put it. You can't be the lead until you've played it, and once you've played it, you have a different feeling from then on. There's a sense of responsibility that goes right up the back of your spine.

I once did *fifty* leading roles in television in twenty-six weeks. How is that possible? A half-hour show had five minutes of commercials, so that left twenty-five minutes of drama. Split that among six people in the show, and the leading man doesn't have an excruciating number of lines. That wasn't nearly as difficult as it was to do the first two-character, hour-long dramatic show on the old "Kraft Showcase." *That* was difficult.

Going back to my "versatility," I did lots of summer stock. I also narrated the first St. Patrick's Day Parade—ad lib. They put me in a museum on Fifth Avenue and dressed me up as a leprechaun. (Laughs.) And when it got very dull with all those guys in black overcoats and green armbands, they'd cut to me and I would sing Irish songs a cappella. I was appearing in *Harvey* at the same time. That's funny, isn't it? There's nothing duller than the St. Patrick's Day Parade, you know. Good thing for my versatility.

Actors like me who survive are not "stars." But we *are* well known.

When I go to England or Ireland, I'm recognized on every airplane, in every damned hotel, every place I go. And I must say, people are getting very polite. Several thanked me for my "body" of work over the years. And some of them could actually name off the shows from theatre, television, and motion pictures.

Now that's very flattering.

Chapter 3 ·

Katherine Helmond

Katherine Helmond talks about achieving success as a mature actress and why that's such an anomaly in American television. She describes how she created the roles she is best known for, and she explains why film tends to be "antiactor" in nature. She also discusses the importance of artistic and intellectual choices in career decisions. Katherine Helmond was interviewed in her dressing room during a break in the shooting schedule for ABC-TV's comedy series, "Who's the Boss?"

. . . .

Career Profile Katherine Helmond has an extensive background in drama and comedy. In theatre, her dramatic work received a Tony Award nomination, a Clarence Derwent Award, a New York Drama Critics Award, and an Obie nomination. She starred in many dramatic plays, including Eugene O'Neill's *Great God Brown* and *House of Blue Leaves.*

In television, she created the role of Jessica Tate in the ABC comedy series, "Soap," which earned her the Golden Globe Award (1980) and four Emmy nominations. Several years later, she starred in the comedy series "Who's the Boss?" and was nominated for a Golden Globe Award (1985).

Ms. Helmond worked for some of Hollywood's finest directors, including Alfred Hitchcock (*Family Plot*), Robert Wise (*The Hindenberg,*) and John Hancock (*Baby Blue Marine*). She worked for Terry Gilliam in England on *Time Bandits* and the highly acclaimed film,

Brazil. She was featured in *Shadey, The Lady in White,* and the Terry Gilliam film, *The Lies of Muenchasen.*

Ms. Helmond grew up in Texas and began her career in local theatres in Houston and Dallas. The announcement of her career goals was received at home "with about as much enthusiasm as if I were selling myself into white slavery." She worked with a variety of repertory companies, eventually forming her own summer stock theatre in up-state New York to produce classic plays. She did ten years in summer stock, and seven with repertory companies around the country.

She was invited to teach acting at Brown University, Carnegie Mellon University, and the American Musical and Dramatic Academy. For new actors she stressed the importance of being "conservative and economical about acting—doing only what is required to make a character believable."

Ms. Helmond added another dimension to her work when she was accepted into the American Film Institutes' Directing Workshop for Women (1983). She's since directed both film and television, including episodes of the comedy series "Benson" and "Who's the Boss?"

For the past ten years, she has been seen regularly on television. She's guest starred on a great many series, as well as in television films. Among the television films are "World War III," "The Autobiography of Miss Jane Pittman," "Wanted: the Sundance Woman," and "The Legend of Lizzie Borden."

She remains very active in all disciplines of the craft.

•   •   •   •   •   •   •

Tell us about your background and how you moved into the craft of acting.

I was an incredibly shy child. Yet, when it came to doing plays, I rather blossomed. I was in high school plays and worked in local theatre in Texas. It was fortunate that Texas had so many theatres, especially in Houston and Dallas. In fact, a group of people who were at the Pasadena Playhouse came back to Texas and formed a theatre company. I worked with them for a while and then went to work in Houston. At that time, Houston had three full-time operating professional theatres.

I didn't have any formal training, and that bothered me. I had on-the-job training, but I considered myself completely *un*trained. I felt I was missing a great deal. The more you learn, the more facile you become, the greater your chances of longevity in a career.

I was in a situation where ignorance was bliss. I just leaped in. When

I was asked if I could do Shakespeare, I said, "Sure." But I had to work on my own speech, because I had a regional accent that wouldn't sit well in Shakespeare. Coming from Texas, I had a flat way of speaking with an "I" and an "E" substitute, which is peculiar to the region. It was the Gulf Coast area, so the speech pattern was not quite as flat as east Texas, or western twangy as west Texas. But it had to be eliminated.

It's always been a great desire to study every aspect of acting. It would have been a tremendous asset to have studied Shakespeare. It would have been wonderful to have a genuine voice coach. Diction and dialect training would have been enormously helpful for jobs. With physical training, I could have learned how to use a fan, pour tea, dance, do classical movement on stage. All of that I had to learn piecemeal while working and watching other actors.

Were you influenced by any other actors? Did you pattern yourself after anyone?
As far as American film stars, I thought that Katherine Hepburn was *the* most wonderful actress in the world, because she had a kind of vibrancy. An English actress that has that, too, is Vanessa Redgrave. She has an iridescent quality of energy that I feel is tremendous on screen.

My all-time hero was Laurence Olivier. I thought he was the consummate actor, because he's been able to work in every arena. He started as a handsome leading man and has gone on to be a character actor. He's played comedy, tragedy. He's made films, acted in television, directed, he's written books. He was an actor who pushed at all the perimeters and expanded through the years. I wanted to be just like that.

When you get a script, how do you analyze it?
The first thing I do is read it in one sweep. I have a fairly visual imagination, so it unfolds in my mind's eye like a movie. I'm looking at it as a viewer. I get the original feeling and sweep that a person would get when seeing the end product in a theatre.

Then I go back and read all the scenes my character is in. I read them, analyze them, see what can be done with them. I try to analyze the function of the character in the overall piece. At that point, I'll go through the script and see what other people say about my character. That gives me hints on what to play.

Before you start rehearsal, do you do any character development on your own?

I work from the hints given by the writer. I'm assuming the director is working off those hints, also. At least it gives me a starting point with the director. Some of those ideas may get changed. Some get dropped, some can't be incorporated, some can be added and expanded.

If I come in prepared, with a notion about character and relationships, then I'll be a more vital force in rehearsal. The director can say, "I think that's off on a wrong track, let's try this," or, "let's explore that and see where we can go."

I think the physical look of the character is very important. How they walk and talk is important, because it makes them unique. Just like people in life are different. I also think about the emotional life of the character, the mental life, and the relationships.

If it's a comedy, I think about the "persona," the personality of the character. In comedy, you need to bring in a three-dimensional persona. That brings a thrust and vitality to comedic acting. If it's a drama, then I dig a little further into the psychological aspect of the character.

The character of Jessica Tate was wife and mother to a strange brood of people in the comedy series, "Soap." How did you develop her?
Jessica Tate was the strongest person in that family. She operated out of love and caring for other people. No matter what came along, it was her family, she loved them. She was standing by them. I remember telling Susan Harris, who created the piece, it would be a mistake to make this woman silly and vapid. It might be funny for a few shows, but it would become obvious, and the audience would tire of her.

I approached Jessica from the fact that she does not operate on the same daily reality that other people do. She's always been isolated, she's always been rich, she doesn't deal in the actual world. Her world is real, but it isn't the actual, daily, get-out-in-the-street, go-to-work world. So her approach and attitudes about things are colored in a different way.

F. Scott Fitzgerald said, "The rich *are* different." And indeed they are. They come from a different point of view and a different reference point in life. I wanted to show that Jessica was very much in her milieu. She was the person who was the Iron Butterfly, but just out of step with the actuality that other people were living in.

The thing that tied her together with the group of people in that setting was that she loved them—even though they did terrible things. She loved them, even though they made bad mistakes. That was her strength in life.

The role was certainly different than the character of Mona in "Who's the Boss?" Can you tell us how you developed her?
I saw this woman much more like my grandmother. Very sure-footed, straightforward, a very "now" woman. My grandmother dealt with what is. "If that's what is, that's what is," she used to say. The common sense of that attitude is incredible. It can get you through life so much easier.

I thought that Mona was very much a woman of her time and of the eighties. She had been a woman who married young in college, had a child, had a husband who provided and kept her at home. Suddenly, all that's taken away from her. And instead of withdrawing, she went *out* into life and said, "Wait a minute. Let me see what life's about. I've still got mine and it's good, and my daughter has problems. My grandson has problems. We'll fix them."

Katherine Helmond as Mona Robinson in the ABC-TV comedy series, "Who's the Boss?" (Courtesy of Columbia Pictures)

What acting techniques have you found most practical? What's your reaction to the Method as a technique for preparation on stage, film, and television?
The Actor's Studio popularized the Method and turned out some of our country's most outstanding actors. I've found them very committed, very serious about their work. But it's quite difficult for them to work with any kind of speed. Many of them spend a great deal of time exploring and preparing for their roles. Sometimes that creates problems, especially in summer stock or movies.

There's just no time for extensive rehearsals and character preparation in film. It doesn't work unless you are *the* star and the script is written collaboratively with you. The reality is there is not much time spent rehearsing movies. The only arena in which one can have that kind of leisure is theatre.

And so Method actors find it very difficult to deliver in the given amount of time. As a matter of fact, in the television series I'm doing now, "Who's the Boss?," sometimes very good actors are let go within a day or two, simply because they can't deliver fast enough.

In television, the script is rewritten everyday. You have to be prepared to let go and move on. You have to be able to make decisions and follow through. You have to come to conclusions about the character. You just aren't allowed the pleasure of exploring it to the "nth" degree.

Let's talk about the rehearsal process. Would you compare the rehearsal process in theatre, film, and television?
In the theatre you run the risk of being over-rehearsed. Tyrone Guthrie cautioned against it, even in the classics, because you lose a certain spontaneity. You lose a sense of danger, where you walk on the edge of do-you-really-have-it-under-your-belt? Is it in your grasp? A vitality can go out of it.

One must rehearse in the theatre to the point that you're secure in what you're trying to do. You're secure in what you want to project in your lines. I have great respect for the author and the text. The full grasp of the text can only come with a certain amount of rehearsal.

There was a time when Robert Mandan and I went to work on a play (*Same Time Next Year*) with only one week to put it together. I was on another project and didn't have time to learn lines. I had seen the play, saw the film, read the play. But, I hadn't learned the lines verbatim. So we really had to leap into it. It was quite wonderful, because Robert's a marvelous comedian, a good and dedicated worker. We like each other, we worked together on "Soap," we were in sync.

Acting is fifty percent reacting. So, if you're really at one with your fellow actor, and you're really watching them, listening to them, fifty percent of the work is done. You have to have incredible trust with the other actor on the stage. That's the key.

Robert and I had that incredible trust on stage. We brought with us that absolute trust the other person was going to know the line, deliver it right, help you get the laugh, be there on stage with you every second. So we were able to pull it together by the end of the week, and we had a terrific time doing it.

We toured the show, and that was our chance to expand and explore our characters. That was our opportunity to go further into the characters and to find little shadings and colorings that we had not had time to find in a week. (Laughs.) We barely managed to get all the costumes and wigs worked out.

Now in film, rehearsal is very different. You come to the set with the words memorized. Their idea of rehearsing is to get you in the light and make sure you hit the right mark for the camera. The first film rehearsal I ever did would have been considered a technical walk-through in theatre. After it was over they said, "Let's shoot it." I asked, "When are we going to rehearse?" The director said, "We just did." I told him that was walking through for technical problems. And he said, "That's all you're going to get, little lady."

From that point on, I realized an actor must come to the film set *very* well prepared. You've got to come in with certain conclusions. You can also run lines with your fellow actor on the set. You can try to talk to the director, but it's not easy because they're preoccupied with other responsibilities.

In television, you've got to be facile. You're up for grabs, because of the rewrites everyday. You simply have to be able to memorize in a hurry, drop ideas, try new ideas. The rewrites are mainly for punching up jokes, getting laughs. As long as there's time, they'll keep changing. You have to be brave. You have to be willing to make a fool of yourself. Willing to fail.

You've directed a lot in recent years. What do you see as the ideal relationship between actor and director?
What I really like is when the actor and director are both going for the same thing. There's input from both sides, and you feel, "Good. We talked it through. We *both* are in agreement."

It doesn't happen often. And it happens more in theatre, simply because of the time factor. Film and television directors are strapped for time. They're also responsible for all the technical aspects of the production, keeping it all under control.

I think the best kind of relationship you can have with a director is one that's mutually enthusiastic and supportive. It's a director's job to be prepared, have a point of view, express it well, and instill confidence in actors. Together they'll find a way to solve problems.

My experience as a director helped me become a more reliable actor, especially in television. In television, actors work so fast. They have to make accommodations to the camera, to the given situation, to the light, to get both eyes in the film, and so on. I've become more responsive and reliable. I know I'll get in the same spot every time during rehearsal. It's reassuring for the director to know actors will be there every time.

You've worked extensively in stage, film, and television. How would you compare the acting media?
I think stage is the actor's medium, film is the director's medium, and television is a writer-producer's medium. Ultimately, television is the producer's, because they have the final cut and the final say about everything.

Directors are very strong in film. They conceive a film and see it through to the end. Even if they don't write it, they see it straight through to completion. It's their project. A Terry Gilliam film is a Terry Gilliam film, and you know it. Nobody can step in and take it over. It's his vision and his concept. It's totally a director's medium.

But, the stage is different. Once the curtain goes up, it's in the actor's hands. You play in a continuous arc. It has a beginning, a middle, and an end, and you have a thrust to the performance in the evening. You have that arc of performance. You're responsible for every word, every movement, every bit of information to be conveyed, every laugh. Some nights are better than other nights. That's what makes it vital and immediate. That's what keeps the actors alive. Audience feedback is immediate.

In television series, you're doing a little movie every week. Before I did television, I used to think, "Why can't they all be wonderful?" Now, after seeing how difficult it is, I marvel that so many of them can be good. It's only dogged determination and professional determination that make them come out well.

When you're doing a film, do you like a lot of takes, or do you prefer to get it done and move on?
I like to do it and move on. But you can't always do that. There are technical things that get in the way. It's not anybody's fault. Clouds come over and ruin the lighting, or the camera simply cannot get all the way around, or someone overstepped their mark. You have to keep doing it over.

There are some actors that like thirty or forty takes. I think of them as "cold" actors. I don't really mean that in a negative way. I mean they start cold and they have to warm up. The more they do, the more the blood flows, the more breath gets in, and by take thirty everything is vital and going.

For me, I find that it flattens my acting a bit. I might have my best performance in take three, and they decide to use another take. If the scene played really well, there is that twinge of apprehension that I won't be able to do it that well again. It's frustrating when you see the film and you feel that the best take was not chosen. *Your* best take. They may have chosen the take in which the lights and camera look best.

On stage, of course, one does those scenes over and over for a year. But, it starts at the beginning, and there's continuity. It has a build-up, a continuity. You have the base from Act I, Scene 1, all the way through. Whereas, in film, you're doing it in such bits and pieces, you just hope that on that line you can get back the same emotion. Or you hope that you'll be able to get just that right twist on the reading of the line to get the laugh.

Do you find it difficult to shoot out of sequence?
Yes. It's very difficult for the actor. I think film is very *anti-actor*. There's nothing there to help you, everything is against you. The stage is pro-actor, because everything is in sequence. The character slowly builds, the jokes slowly build, and the drama slowly builds. You have it in your hands at all times.

Out-of-sequence shooting is very confusing. And that's where you need an outstanding director to remind you "Six weeks ago, the three lines before this, remember you felt like this?" The director has seen the dailies, seen your performance over and over, and has a better memory of what you did in that scene.

Dailies are set up to show the previous day's work. Do you think actors should watch their work in dailies?
That depends entirely on the actor. I learned a great deal about film acting by watching dailies. I think it's very helpful. For actors that are highly neurotic about their work, I would say no. If they're too concerned about their looks, they should *not* watch dailies. It interferes with what they'll do the next day. They're very upset about hair, or clothes, or something.

I'm an actor that never worried about looks. If anything, I'm pleased that they've pulled me together. If the clothes are right, the hair is right, then it's right for the character.

But, what I *do* see is the emotional quality or delivery of a line that's supposed to get a laugh. I can see if I'm totally in the scene or not quite there. I see a lot. And I see a lot by watching other actors and how they do things.

What do you look for when you select a role?
I look for a character that I can identify with, by humor, or by drama in their life. That doesn't mean that I've committed a murder, or lost a child, or was locked up behind the Berlin Wall. I may not have personally experienced those things. But I can identify with the emotional or the humorous content of the character. That's when acting is easy, if you get that kind of part. Then you can sail right through it, because there's a little kernel of identification in there for you.

Has typecasting been a problem for you?
Yes. Before "Soap" I did years of drama, and I couldn't get a reading for a comedy. They considered me a straight dramatic actress. Then, after "Soap," they only gave me comedies to look at. Now, after having two hit comedy series, with extreme comic characters, I think people are asking, "Can she be serious? Can she understand drama?"

It's maddening. You can either become angry and bitter, or you can just keep pushing.

What are your thoughts about casting and the audition process?
The casting process is a problem because it doesn't really reflect an actor's quality. Some people give wonderful performances in the first reading, but can't do anything beyond that. Then you see people who are terrible in the audition process, but if they get the part, they become much more secure.

I never thought I had the ability to audition. If I just got the part, I could do it. I really wasn't equipped to audition. Somehow, coming to Hollywood freed me up. I didn't take the process too seriously, because the interviews were all very casual, social, and funny. I didn't even have to read for parts. People brought actors in to talk, to see your "quality." That's what's important on film. That's freed me up.

Your success as a mature actor was attained over a long period of time. What are your thoughts about achieving success at an older age, and maintaining longevity in this career?

Ultimately, I think it's better to have success grow slowly. It can sustain you till the end of your years.

Some young people come into television with little or no training. They get to be superstars. They make more money than they ever dreamed possible. They have an entourage of people "yessing" them. None of them are encouraged to go to classes or summer stock or grow as an actor. It's all just, "get the next job."

Then, one day the TV series is over, and their careers are gone. They can't understand what went wrong, because they did everything they were told to do. It's very hard for those young people to ever rev back up.

Compare that to people my age, like Robert Guillaume or Bob Mandan. When we came into "Soap," it was practically the first TV we'd ever done. We've kicked around so many years, done so much summer stock, and repertory, and soap operas, that the increments by which we went forward were so slight. We felt steady—as people and performers. We were not likely to be overly flattered or swept away by the fact that we didn't have to stand in the unemployment line anymore.

Because we know it can all disappear too. There was a line Tennessee Williams wrote in a play that I always equated to acting: "Well, you can be young and poor, but you can't be old and poor. It's ugly." (Laughs.)

I feel quite fortunate that I'm still very actively working at my age, not that I'm an antique. I'm beyond thirty, and that's usually when the cutoff begins in American television and film. I hope to stay very active in the field of entertainment to the very end, because I think that's what keeps me young and vital.

It's miraculous if you can maintain any kind of longevity in a career in the United States. We don't have that respect for mature actors. Ruth Gordon was one of the few actresses that worked all her life, into her mid-eighties. She acted in a movie, went home, went to sleep. She also worked as a writer, producer, and author. She kept a vitality in her life. It was amazing. She had great wit and professional ability. She could deliver a line that could knock you out. She had determination. She had to act. All her life.

In Europe, you see much more respect for that kind of longevity. Some actors last well beyond their eighties, and they're still working in films.

Recently, I received a letter from Arthur Peterson, who was in "Soap," played the crazy grandfather. He and his wife, Norma, have been in show business and married for fifty years. He's out touring a play, she's doing voices on television. They support a writer's theatre. They're very happy. They're artists. They could never do anything else.

That's why "Soap" was so wonderful for them, because it gave them a little cushion for their old age. With his education, Arthur could have been a teacher, a college president. But he'd say, "No, I couldn't be anything but an actor." That's exactly how I feel.

How do you deal with the insecurity of the business of acting? After several years of doing "Soap," for instance, it was suddenly pulled out from under you.

Indeed it was. After four years of bouncing along successfully, we left to go on hiatus. I did a "Loveboat" special in Australia and was told the series was coming back for the fifth year. While I was in Australia, somebody told me that "Soap" was going to be dropped. That was quite a shock.

I didn't do another series on the air for four years after that. During those years, I did four or five pilots, about a pilot a year. When I did the pilot for "Who's the Boss," I actually had done three pilots that one year.

So, you start all over again. You're an apprentice all over. You're starting at point zero. The only thing you bring with you is your reputation. And, by virtue of your work, the entry into the readings is more available. But you're out there slugging away, trying to find the next job. Trying to deal with the unknown factors in each job you take.

If you need security in your life, you must *not* go into acting or into the entertainment business. Because it is mercurial. We know that there have been stars for twenty-five years earning top money and box office draw that have not worked in ten years.

I stay with it because I love acting. I'm most fulfilled when I'm working. And so, when I get a taste of it, even when the stretches in between are long, it can hold my spirits up until the next job.

What do you think there is about your personality, about you, that enabled you to beat the odds and become a success.

Stubbornness, more than anything. If I say I'm going to do something, I'm going to do it. No matter what comes up, if it's difficult or inconvenient, I'll do it.

I have my grandmother to thank for that. That's how she raised me. My mother was only eighteen when I was born, and she went back out to work. My grandmother lived with us and made us stick to commitments. If you said you were going to do something, you did it. I joined the drum and bugle corps and the first time we had to march out in the sun, I said, "Oh boy, I'm not going back to do that." My grandmother said, "Oh yes you are." So you think twice before you give your word. A person is their word. If your word's no good, you're no good.

So, I had a streak of stubbornness to say, "This is what I'm going to do with my life," and a commitment to see it through. I wasn't a good hustler, wasn't a good auditioner, but people gave me jobs, just enough to keep me hooked in. The steps up were very slow, and I almost despaired many times.

There were times I was working in the theatre and had two other jobs. I worked all day long teaching and did eight shows a week off Broadway. I had a day-off job, just to pay the rent. But, those were the good times because I was also in a play. I loved acting.

And there were times I didn't work for a whole year. I'd be in a typing pool in an office, or working the switchboard, or running errands. I still was stubborn. I wanted to act.

I am happiest acting. I am the most fulfilled when I am acting. It's given me the most reward, the most joy in life.

What advice would you give to new actors?
If you can make a living doing anything else, and have a decent life, that's the choice you should make. If you feel you *must* act, that's another story. If you can't be anything but an actor, you will find a way to be one.

Parents shouldn't discourage their kids from acting simply because the money is erratic. Not everybody can be rich and famous. Even lawyers aren't guaranteed wealth, especially public defenders. They are down in East Los Angeles in the barrio. They'll never make any money, and their parents spent a fortune getting them to be attorneys.

There are doctors working in the boondocks because they want to help poor people. They've spent hundreds of thousands of dollars on education to become a doctor, and they're driving old cars to go out and deliver babies.

Mel Shapiro, who was head of drama at Carnegie Mellon University, has been in theatre all of his life. His two children decided not to go into the theatre. He said, "Good, that's wonderful. I'll give you the best education." Well, of course they both turned out to be absolute geniuses. They're interested in philosophy. They're going to be philosophy teachers. He said, "Good luck. I'll have to send their kids to school." And I said, "You can't ever tell what people will choose in life." But if it makes them happy, they found the right choice.

What could make a parent happier than to have children choose something that will give them a sense of accomplishment? It gives them personal reward. Artistic and intellectual satisfaction are worth an awful lot. A lot of young people go for the bucks. Then at thirty-seven, they've got money, they've got the house, they've done everything, and they're not happy. And they wonder why.

It's wonderful to choose what you want to do with your life's work and pursue it to the end. Choice is the greatest gift one can have in life, about your career or about your life. As you grow older, you see how few people in the world do have choices—in the work they choose, where they live, what they do in life.

That's one of the reasons I've grown more grateful that I have been able to work as an actor. It's a thrilling way to make a living, and I feel very privileged. I'm very catered to in my daily work arena. I know I make people laugh at the end of the day, and that's very pleasant. That's the nicest thing about being a successful working actor.

Chapter **4** ·

Robert Mandan

Robert Mandan talks about his early studies in the Method and how Stella Adler's techniques helped him significantly. He discusses the problems of typecasting, and comedy style, in television. He also describes an unusual technique used by film actors in close-up acting. He presents his views on the feasibility of a national repertory company and provides sound advice about training for the craft of performance. Robert Mandan was interviewed at a restaurant in Studio City.

· · · · · · ·

Career Profile   Robert Mandan has performed in television, film, and theatre, with a wide variety of roles, ranging from Shakespeare to television comedy.

He studied drama at Pomona College, joined the Palm Springs Playhouse, then worked in summer stock. In New York, he performed at the Equity Library Theatre, completed his degree at New York University, and studied acting under Lee Strasberg, Herbert Berghoff, and Stella Adler. He did his share of ''out of work actor jobs,'' including typing for the National Biscuit Company (he claimed to be writing for NBC).

In New York, Mr. Mandan appeared on stage in Sartre's *No Exit,* and award winning productions of *The Trojan Women, Julius Caesar,* and *King Lear.* He also appeared in daytime television dramas ''From These Roots,'' ''As the World Turns,'' and ''The Edge of Night.''

He guest starred in live television productions of "Kraft Television Theatre," "The U.S. Steel Hour," and the last live telecast of "Armstrong Circle Theatre."

Mr. Mandan played the role of Sam Reynolds in the daytime television soap opera "Search for Tomorrow" for three years (1967–1970). He appeared on Broadway in *Maggie Flynn, There's A Girl in My Soup,* and *But Seriously,* which opened and closed the same night. His next Broadway role was more successful, starring in the long-running Tony award winning musical, *Applause.*

After moving to Los Angeles, Mr. Mandan guest starred regularly on primetime television in episodes of "Mannix," "Petrocelli," "The F.B.I.," "Cannon," "Streets of San Francisco," "Barnaby Jones," and "The Manhunter." He starred in "Caribe," then returned to episodic TV work in hundreds of shows, including "Marcus Welby, M.D.," "Sarah," "Delvecchio," "Kojak," "The Rockford Files," "Police Story," and "Medical Center."

In the ABC comedy series "Soap," he starred as Chester Tate for four years (1977–1981). During that time, he continued to work on stage, and he starred in the CBS-TV remake, "You Can't Take It With You." He was featured in the comedy series "Private Benjamin," and co-starred with John Ritter in the comedy series, "Three's a Crowd" (1984), a spinoff of "Three's Company." He's played major roles in motion pictures such as *The Best Little Whorehouse in Texas* and *Zapped.*

Mr. Mandan continues to perform regularly in television, film, and stage.

．　．　．　．　．　．

You started out as a theatre student in Los Angeles and went to New York to be a stage actor. Can you tell us about that experience?
I went to Pomona College, a small liberal arts school in suburban Los Angeles. I studied under a wonderful woman—Virginia Princehouse Allen. If she had not locked herself in this small school, she would have been a major force in the American theatre, because she was a brilliant director. She had gone to Yale drama school and then came right back, started the department and ran it single handedly. She was an incredible producer and director.

One of the things she taught us was to trust your intuition. She assumed that there was an intelligence at work that could read the script, analyze it successfully, and come to a conclusion about the text.

When you were in New York you studied the Method?
Yes, I did what every young actor does in New York. At the time, the
Method was the only game in town. Everyone was teaching and studying the Method, but I just couldn't get with it. I felt terrible. I studied
with Lee Strasberg, Herbert Berghoff, Gene Frankel. They were all
doing the Method, but it was just totally wrong for me psychologically, technically, artistically. They insisted on digging into emotions,
private moments. I wasn't in any condition—as a person or as an
actor—to bear my soul like that.

While struggling with this, I ran into a gal who was an absolute
devotee of Stella Adler's. She suggested I study with Stella, but I
thought Stella was a crazy woman; I wasn't about to put my professional life in the hands of a mad woman. This was in the 1950s. Flash-
forward. In the 1970s, I was back in New York. Stella was offering a
summer course, and I thought, it could be fun. I haven't taken class in
a long time.

After the first session, I came home and cried my eyes out. I realized
I wasted all those years! I should have studied with her from the beginning. She offered a direct extension of the approach that I had been
learning in school. She taught us to approach emotions through the
material, through the text. She didn't probe and pull psychological
strings.

*Stella Adler's approach emphasized material in the script rather than
personal emotions?*
Exactly. It doesn't matter whether it's Shakespeare, Miller, Williams,
Inge, Ibsen. You first go through the material and let your feelings
follow you into the construct you build out of your own emotional
pool. It's not a direct route to emotions. It's just another way around
it. You can fool yourself and think that you're not really dealing with
your own agendas, but you are, of course.

Stella teaches "there is no art without imagination." She sparked the
imagination of some of our best actors. She trained Marlon Brando,
Robert DeNiro, Al Pacino. Afterwards they went to the Actor's Studio,
which mistakenly got the credit for their training. Lee Strasberg got the
credit, too, though Stella actually trained them.

Did that create conflict between Stella Adler and Lee Strasberg?
Apparently, there was a great deal of conflict between the two, and it
sometimes spilled out into classes. I was in Stella's class a few times
when some of Lee's students would say things that were obviously

"party" line: "Raw emotions or it's false," that sort of thing. She would annihilate them for it. She had no patience for delving into personal feelings, especially if the script doesn't call for that approach.

Stella never denigrated the act of being an actor, but she always put the writer first. The writer is the *primary* artist, the actor is the *interpretive* artist. Through the actor's imagination, the material is illuminated. And you *always* worked from that material. She didn't care how, or what you were feeling, or what you've ever felt. "If you couldn't illuminate the material," she used to say, "who cares?"

She was always battling Strasberg over the "correct" interpretation of the Stanislavski system.

Stella Adler actually worked with Stanislavski in Paris. Did she discuss that with you?
Yes. She went to Paris to meet Stanislavski and work with him. Meanwhile, Lee Strasberg was teaching here, not really using the "real" Stanislavski system. It was Stella Adler who actually met him and worked under his tutelage.

She told a story in class about something Stanislavski did. It was an "inner life" situation. She was supposed to go to the window and look out. She had this whole "inner life" planned, going, moving for herself. But, every time she went to the window, he'd say, "I don't believe that." She couldn't comprehend it, because she was so "full" and so "living" as she was crossing the stage. Finally, he stopped her and said, "Okay, tell me what's outside the window." She said, "Well, I don't know." And he said, "Perhaps that's why I don't believe you."

With all this inner work she had done, she had never put anything outside the window. Her point was that the set designer can put something there, but *you* must imbue it. Everybody teaches that you have to "imbue" that neon light, that telephone pole, that vista, whatever it is. You have to "imbue" it with a life that's meaningful to you.

I always loved the way she taught because she taught in degrees. You could "imbue" that with "light," "medium," and "dark," as she called it. Stella says you can do anything, as long as you keep in mind those degrees of intensity. Value. Whatever word you want to attach to it.

I think a lot of young actors have difficulty with this concept because they work from their own emotional life, which often times is rather dark. Many young actors, I think, have come from rather heavy backgrounds emotionally. They want to fill their acting with that. But that provides a "sameness" to the acting, and it may be inappropriate to the play or the situation. It disenables them to do light comedy, for instance.

Sir Laurence Olivier in his book On Acting *refers to a legendary line of classical actors who influenced him.*[1] *Who influenced you?*
The two biggest influences on me were Sir Laurence Olivier and George C. Scott. I first saw Laurence Olivier on stage in *The Entertainer* in New York, and I thought I was going to go mad. Some other actors felt he was cold, there was nothing really going on inside him. I didn't know if that was true. But he sure had me fooled. In the audience, that's all I cared about. I didn't care how he did it, he moved me to believe in his character. He was a great craftsman.

All the stories I've ever heard about him have amused me and inspired me. One I call "the green umbrella story." He was doing a part and it didn't come together yet. He had done everything. He had gotten the voice, the walk, he knew the emotional life of the character. But, it just didn't fall into place. It suddenly came to him one day to carry a green umbrella.

Now, that's pure inspiration! How does a thought like that drop into a brain? They found him a green umbrella, and the minute they put it into his hand, the characterization fell into place for him. It somehow focused everything that he had done before.

I'm particularly fond of that story because I had a similar experience, but with no success. I was doing the role of Don Juan in hell. It was essentially a literary character, and making it live on the stage was very difficult. I tried several things, eventually settling in on a Spanish accent. I must have sounded like a Mexican orange picker. The director asked me kindly to discard it, and I gave in. I didn't realize that I actually needed something like that to help stay on track.

We opened to terrible reviews, played to five or six people a night; it was painful. The director returned from a business trip to London, saw what was happening, and was desperate to put new life into the play. She urged me to try the Spanish accent again. Actually, once I did, I realized it could have worked. It gave me an "exotic" nature. The problem was I couldn't just fall into a perfected Spanish accent overnight. We closed in a few days, but the lesson stayed with me.

. . . .

[1] Richard Burbage, David Garrick, Edmund Kean, Henry Irving are noted as four actors who handed on the Shakespearean mantle to our generation. See Laurence Olivier, *On Acting* (New York: Simon and Schuster, 1986), 35.

I always think of Olivier's story. He was secure enough to stick to his guns, to know the track he needed to be on. Olivier always does incredible things with his characters. He's done some *spectacular* things on stage. Wasn't it in *Coriolanus* that he leapt off a parapet tied by the ankles and just swung there? He's so very athletic and physical.

Olivier also has a sense of humor imbued in his roles. He always managed to get something funny into a performance. He'd make fun of somebody in an effeminate way, or do something that would be a twist from the character. In *Richard III,* he played the man with such intellect that he could do all this "ornamenting" around the edges. He understood exactly the way people really behave.

I always found that refreshing—an actor with a sense of humor. Olivier had a sense of humor about himself, about his role, about the play, about the whole scope of the theatre, the world. It's very endearing.

I mentioned George C. Scott as another influence. He's one of America's best actors. Over the years, he has consistently turned in totally believable performances. I've never seen him in a role I didn't fully believe.

The first time I saw him hit the stage in New York, I knew he would change the complexion of American acting. He's always had this incredible talent to be somebody else, to take on a completely credible persona.

What process do you go through when you get a new script? Do you research characters and situations?
Unfortunately, in television, you can't really do that. Most of the work isn't that rich, doesn't have that much substance to it. And there is another unhappy reality in television and film acting, called *typecasting.* If you have certain success in a role, that's what you'll be offered. In that case, you know, intuitively, what the role is all about.

If you're lucky enough to have a role that's different from you, *then* you have to do investigative work, imaginatively. I did a play several years ago, called *Girlie, Girlie and the Real McCoy.* It was about an old burlesque vaudevillian and his troupe, which included his sidekick comic, a straight man, and a new young burlesque stripper.

I wasn't familiar with that particular milieu, but the director was. He spent a lot of time down at burlesque houses just with the intent of writing this play. So, a lot of the jokes, a lot of the routines, he knew and brought to us.

The basic emotional life of the character, however, you have to investigate on your own. Who is he? Why is he? What is he? How does he talk? How does he walk? How does he feel? What does he think? How does he feel about each and everyone else in that play? How does he feel about where he lives? How does he come to work in the morning? How does he get there? What happens to him on the way? The richer you can make all that, the better.

I keep quoting Stella Adler, because she encapsulates so much knowledge about acting. She said you don't have to spend hours and hours and hours discovering that stuff. You make choices. And if one choice doesn't work, you make another choice right on the spot.

She used to do exercises with us, changing our attack on something in midstream. She'd make us choose something else that might have more weight to it, less weight, more color, more profundity. It's imaginative and fun. Then you're not afraid to say, "I'll try this, I'll try that."

You did the comedy series "Soap" for four years. How did you keep the character fresh?
The character was always being put in some new situation. There was not much character development or change over the four years. The changes came in the situations. The character was pretty much the same at the end of four years as at the beginning. That's why I got cast. I was very much stereotyped in that.

How do you react to that kind of stereotyping in casting?
To this day, more than a decade after "Soap," I'm still fighting stereotypes. I've been turning down stuffy, Waspy, father roles in innumerable scripts. I've turned down two or three features, because I'd read three pages and I knew the role they were submitting me for. The script's storyline is inevitably about a girl's father who's a crass executive, and a young boy has to get past him to get the girl.

It's invariable. There must have been ten scripts that were sent to me that had the exact same role. I finally found the courage to just say, "No. I won't do that anymore."

Typecasting is a real problem for actors. Once you've made it, the audience doesn't want to see you vary too far from what they love you for. Edward Arnold always played Diamond Jim in his film roles. Walter Brennan was *always* Walter Brennan. Think of some of our biggest film actors, from John Wayne to Cary Grant, the audience *expects* to see them as "themselves."

Interestingly, typecasting isn't a problem in England. English actors have the opportunity to explore a wide variety of roles, and the audiences flock to see them. That's what you were talking about earlier with Olivier feeling part of that lineage of actors. We don't have that history here.

A few years ago, I met Roger Reese in London. When he finished a demanding role in *Nicholas Nickleby,* he went back to the Royal Shakespearean Company, on a three-year contract. That's like going home with extra credit for working out and playing. We have nothing like that in this country.

Would you like to see a national repertory company for actors in this country?
It would be wonderful if we could have a national repertory for actors, but our geography almost dictates against it. Where would it be located? East Coast? West Coast? Midwest?

In England, everything is centralized within an hour of London—theatre, film, television. Even Stratford is an hour and a half from London, tops. It's conceivable an actor can play at Stratford, go to an important film appointment in London, and make it back to Stratford for the following night's performance.

We've probably come as close to it as we can with different companies like the Seattle Repertory and the American Conservatory Theatre. There's also Louisville, Houston, Washington, Boston. But, right now, it's awfully diverse. There is no real centrality.

It's interesting to imagine that there could be a production company located someplace like New York or Washington that would bring in a regional production each year. Let's say it did three or four regional productions a year. This year, Seattle gets it. The next year, San Francisco gets it. That way there could be a central focal point for American actors.

You've worked on stage and television. It sounds like you prefer working in theatre.
Theatre keeps you in touch with audiences, keeps you alive and vibrant. I'd love to do more theatre. There is much more time for actors and directors to work together. That integration of creative visions can merge for a powerful experience.

Let me give you an example. The best production of *Hamlet* I've ever seen was a stage version done by the American Conservatory Theatre. It was staged brilliantly. I don't know how they did it, but it was the perfect merger of acting and directing talent.

I found myself wanting to know who had the poison, who was next going to be inflicted with it. And the duel scenes were brilliant. When it was over, bodies were lying about the stage, and the pages came in to cover them. It was a national tragedy that happened before my eyes. It was just incredible.

My point is that *Hamlet* has been a great play for three hundred years. What's done with it—by actors and directors—is something else. If that whole ambience of acting, directing, and scenic design works, it can be a wonderful experience for artists and audiences.

You've done a lot of work in television comedy. How would you differentiate character work in comedy and drama?
The situation comedy is a much more superficial form. By the time you finish the pilot your characters are already set, your relationships are set. From then on, you *are* doing exactly what the name of the genre says, "situation comedy." The characters are in a new situation each week, to which they react.

The plots are not terribly complicated in sitcoms. You don't have that much territory to cover. In the comedy series "Three's A Crowd," there would be one major plot conflict, or a plot and a subplot. That's all you'd have to resolve in twenty-three minutes.

You can do a *lot* in twenty-three minutes, if you're clever. Despite its thinness, "Three's A Crowd" provided some funny moments over plot crises as simple as misidentification. You can get a lot out of a star like John Ritter. He's a brilliant physical comedian. He would fall over sofas, walk through doors, do all that stuff. The critics slighted it, but it ran for ten years, and the public loved it.

It's actually "low comedy," "pit comedy." It's stuff that Shakespeare used to keep his people's attention during the play. Think of the night watchman in Macbeth; that's a low comic's turn. The audience always loved it.

With rehearsal time so limited in television, how do you prepare for your first day of shooting?
When I started doing television comedy, I went to our first read through, and they'd start timing it. Can you imagine? They were timing a *rehearsal!* That's why a lot of television isn't funny. They'd do an hour show that was supposed to be comedy. By the time it hit the air, it was chopped to nothing. All that was left was the bare bones of the plot and the structure. All the color around it—gone. You look at it, "not funny, boring." That can be a real problem in television.

You can't really prepare. The difficulty faced by actors is that you rehearse *on the set.* It's difficult when you have to go in and rehearse on film. And sometimes they shoot when you're not nearly ready to go. You end up doing thirty, forty, or fifty takes. That's unacceptable, of course, particularly because of economics. You can't do that many takes unless your budget is in the thirty or forty million dollar range.

There are some cases in which a reasonable rehearsal period is set aside, but that's not the norm. The Barbra Streisand film, *Nuts,* based on the courtroom play, had a month's rehearsal. The actors literally rehearsed it like a play. That made the filming much more efficient. Everybody knew what was going on and why.

What do you think is the ideal actor-director relationship?
It's when the director has an overall view of the form and expression of the piece, but allows the actors to find their characters in their own way. It's like designing a new house and letting the actors decorate the rooms. It's okay if you chose blue instead of green. Truth is, some directors can't handle it if you want to change colors.

The pragmatics and economics of film are so pressured, directors sometimes get "wacky." They don't *think* they have time to work with actors' choices. But choices are really instantaneous. If an actor says, "let me try this," the director should give it a shot. More often than not, if the actor has an intuitive sense, it will work fine.

It's a problem if you have a director who is rigid about the form of your character, the form of the play. That inflexibility is especially pronounced in television.

I remember doing a television show, and I thought my character could use a nice looking cane. I had played this role forty times on television, a white collar criminal, and thought a cane might add something new. Maybe the character was wounded in the war. It was like trying to pass an act of Congress. They saw me in other roles without the cane, any change was a major event, it needed the approval of the studio head. Finally, I found a phrase that worked, told the director, "it would make it easier for me." And he replied, "Oh, okay."

Television is the worst offender against creative flexibility in rehearsal and production. I think film is less so. Fortunately, in theatre, that rarely happens.

In television and film production, actors need to be concerned about camera angles and close-ups. Any thoughts about techniques for close-up acting?

When the director is shooting close-ups, everything's got to be internal, you can't do much physically. I've always intuitively understood that. Some of the acting schools for contract players at the studios used to teach this technique. To this day, I see some of the "techniques" from those contract player days.

I remember rehearsing a scene with a man who was trained like that. We rehearsed until it was rote. Then we sat down to do the first shot. It was an over-the-shoulder shot of him. In the scene, he comes into my office and sits down, and I'm supposed to chew him out. At that point, the only way to describe it is to say he *glowed* at me. Something just turned on inside him. Like a light turned on. It was a very strange thing. But it worked on film, because everything came quietly from inside and just shone up through his eyes.

Sharon Gless, who's very successful on "Cagney and Lacey," does that a lot in close-ups. She'll turn and just *glow* at the camera. Nothing happens on her face. She won't do anything. This light comes on, and it's there in her eyes.

Olivier is a master of close-up acting. They always managed to get a close-up of his eyes. I've never seen anybody that can make their eyes absolutely dead, no expression at all. He played the character of Archie Rice in *The Entertainer*. His line was, "Look at me, my eyes are dead." The camera tilts up, and my God, they are. His eyes were dead.

Film production requires much more quiet, internal work than theatre. It's necessary, especially since the camera is right in your face. I think that's why the great stars really didn't do much. They *seemed* to because they had such presence on the screen. But Gary Cooper, John Wayne, Cary Grant didn't do a lot. There was not a lot of activity *around* them.

What are you looking for in your next project?
When I read a script, there's an intuitive feeling, something clicks. One play, *Girlie, Girlie and the Real McCoy* hooked me emotionally. It had a real pull to it. It also had an incredible role. The character did everything on stage, he was funny, sad, tragic, even died on stage. And the play was *about* him. That's very attractive for an actor, particularly a "character actor" like myself. That kind of role doesn't happen very often.

When actors are between roles it can be anxiety provoking. How do you handle the insecurities of the profession?
I've been fortunate. Even when I was a kid, I fell right into a stock

company, then went to New York where I got into a play. I had some lean periods in New York, that's for sure. I didn't have a stellar career, but I had a very substantial working career. I've always been a working actor. My whole life, I've been out of work maybe fifteen months, except when I've chosen not to work.

For me, the key to survival during those lean times was taking class. If I went without work for a few months, I took class. As soon as I'd start, there would be a job offer. That's the way it is.

Actors should remember that class is the key to beating unemployment. It's always there for you. It keeps you active. Work out at a gym, take dance classes, take voice lessons, take acting classes. It keeps the juices flowing.

What advice would you give to actors about training for the craft?
An actor should have a college education, but you don't have to major in theatre. Try other majors—literature, philosophy, religion, history. *Anything* but theatre. If you're driven toward a career in acting, you'll probably spend the same amount of time in performance, whether or not your major is theatre. Another major will force you to broaden your horizons.

You can put two hundred hours a semester backstage. I know a lot of professionals who went through that. Some of them were smart enough to have other majors. I wasn't. I majored in theatre arts. It's not a significant regret, but another discipline would have forced me to explore other areas.

After college, you can take classes with Stella Adler, Herbert Berghoff, or whoever is the new artistic leader at the time. Or, you can go to graduate school. So why not have a broader base, a stronger foundation when you go there? It can only enhance your life and your career and your imagination.

I still think New York is the Mecca for professional actors to train. With the exception of a few teachers who've come from New York, acting classes in Los Angeles are just vague notions of what film acting should be.

I know one teacher in Los Angeles who has never been on a stage. Matter of fact, he's never had a job. He's never done anything. He thinks he's teaching acting, but it's really an emotional process, a "new age" process. It allows people to have an experience that might be translatable to some film role.

You've offered advice to many people interested in acting and film acting. Can you capsulize your views for us?

If someone is interested in acting as a career, the first thing I ask is, "Do you want to be an actor, or do you want to be a movie star?" And if they pause, I know they want to be a movie star. If they say, "I want to be an actor," the response is simple. Go to New York or London, and study where English speaking actors study. Don't wait for a break in a small town. Because, even though you may get the break, you won't be ready.

If they want to be a "movie star," I tell them to learn acting first, so they'll have a craft. *Most* of all, they need to be objective and hard-nosed about their own attributes. If you want to be a movie star, and you aren't gorgeous, then what *are* you? Are you funny? Are you dramatic? What are you? You have to be able to tell whomever you encounter about that uniqueness, so there's no doubt in anybody's mind about what you are.

If you're beautiful—male or female—the doors open a lot quicker for you. If you're beautiful *and* talented, then your chances at stardom are enhanced even more. The odds may be better, but there's still no guarantee of success.

Tom Selleck, who certainly is a handsome man, did seven pilots before he landed the lead role in the action series, "Magnum, P.I." None of those pilots sold. He looked the same, was the same, acted the same. After a slew of failures, the part and the man finally came together. Now he's been trying to translate that into a successful motion picture persona.

Even if you make it in television or film, I think it's essential to learn acting on the stage. That's what Richard Chamberlain did. He was just behind me in college, and he majored in art. After the army, he started taking acting classes. He wasn't much of an actor, not in school or afterwards. All you have to do is go back and look at the old television series he starred in, "Doctor Kildare." He played the leading role acceptably.

But he did a remarkable thing. Instead of staying mediocre, or quitting, he took his celebrity and went to England. He was trained in Shakespeare, played an acclaimed Hamlet, and returned to the U.S. with a highly respected identity. He's been in qualitative roles since, and has had a very substantial television career. He'd been well advised in his career moves.

Any parting advice for our readers?
Yes. Sometimes young actors waste their energies worrying about the wrong things. Some of them worry enormously about gaining a union card—from Actor's Equity, Screen Actor's Guild, American Federation

of TV and Radio Artists. I tell them to forget it, just concentrate on being trained correctly and looking for work wherever you can.

I did a seminar a few years ago in South Carolina, and a young boy came up, asked if I remembered him. I didn't. He said he talked to me in Dayton where I had done a tour. During that time, he wanted to know how to get his Equity card. I told him to forget it. "You don't need your Equity card. Just go find work." His emphasis was all in the wrong place. He was working to get a union card and ignoring his craft, art, and talent.

I gave him that advice years ago, and here he was now. He told me he was resolute about looking for jobs, worked wherever it took him. Then he pulled out his wallet and told me it all paid off. He now had an Equity card, a SAG card, and an agent.

I offer you that same advice. If you have a serious interest in acting, train hard so you know the craft, look for work wherever you can, and do it.

Given the odds against success, what do you think it is that helped you succeed?
Talent. I'm finally able to say that about myself. I've always been much too modest. I'd shrug and say, "Well, it's just luck." After all these years, I don't really believe in luck. I do believe in being in the right place at the right time, but if you're at that right place and time, and have *no* talent, what then?

The bottom line is talent. I don't necessarily mean great acting talent. There's a talent to being a celebrity. There's a talent to being a movie star. There's a talent to being a public persona. If you've got that talent, and it meshes with the right time and place, look out.

Chapter **5** ·

Lynne Moody

Lynne Moody talks about researching roles and analyzes the process she went through for the miniseries, "Roots," in which her character aged sixty years. She also talks candidly about the unexpected year of unemployment following that telecast. She relates the unique and challenging aspects of joining a successful primetime series, "Knots Landing," as a newcomer. Lynne Moody was interviewed in her home in the Hollywood Hills.

. . . .

Career Profile Lynne Moody has starred in television series, miniseries, and films. At a time when few black actresses were able to make it in primetime television, she won the role of Alex Haley's great grandmother in the critically acclaimed TV miniseries "Roots" (1977). A decade later, she was signed as a co-star in the CBS-TV primetime series, "Knots Landing."

Ms. Moody grew up in Evanston, Illinois. Her father was a doctor, her mother a social worker; her aunt was the first black delegate to the United Nations and the first black woman judge elected to the state of Illinois.

Lynne Moody studied acting at the Pasadena Playhouse in Los Angeles, the Goodman Theatre in Chicago, and the University of Washington. After a stint as an airline stewardess, she decided to "sink or swim" in the acting profession.

Her first job was as a Playboy bunny in 1972, then she won a leading role in the ABC comedy series, "That's My Mama." Ms. Moody appeared in a movie of the week for ABC, "Nightmare in Badham County," then won the demanding role of Irene in the history making ABC miniseries, "Roots." She was signed to play that same role in the sequel, "Roots, II."

After the critical success of "Roots," she was dismayed to face a year of unemployment. She planned to move to Seattle, where she was accepted into law school. Prior to leaving Los Angeles, she auditioned for the ABC-TV comedy series "Soap," and won the part of Polly.

Ms. Moody starred in that comedy series for a season, then guest starred in a number of dramatic episodic television shows, including "Magnum, P.I.," "Hill Street Blues," and "Lou Grant." She also attended the first year of Robert Redford's Sundance Institute in Utah. In motion pictures, she played opposite Richard Pryor in *Some Kind of Hero* and was featured in Paramount Pictures' *White Dog*.

In the early 1980s, she was cast in several pilots, but none made it to the network until "E.R." (1985), with Elliot Gould. The following year was a particularly busy time. Ms. Moody co-starred in two television films, NBC-TV's "Society's Child," with Philip Michael Thomas, and CBS-TV's "Lost in London," with Ben Vereen and Emmanuel Lewis. She also guest starred in a number of television series.

In 1987, the producers of "Knots Landing" were looking for a talented black woman to join their cast in its ninth season. Ms. Moody was offered the role of Patricia Williams and has appeared regularly in that primetime soap.

• • • • • • •

You've studied acting for years. Can you tell us about your training and background?
I wanted to be an actress since I was three. All my classes in school were related to theatre and the performing arts—drama, dance, piano. I went to Pasadena Playhouse, which was an accredited college of theatre arts in Los Angeles. I also studied acting at the Goodman Theatre in Chicago, then went to the University of Washington in Seattle for my academic training.

I never wanted to do anything but act, and there were no black actresses, no role models for me. Absolutely none. When I was at Pasadena Playhouse in my freshman year, they asked who influenced us. I only knew of one black actress who was appearing in the theatre, Ethel Waters. She was appearing in *Member of the Wedding*.

Most of the students didn't even know who she was. They looked at me very strange. Ethel Waters was playing a maid, so it wasn't necessarily a well-known part. In retrospect, it wasn't an exciting choice, but she was the only one.

I have powerful women in my family who were successful. They influenced me to shoot for the moon. My mom definitely instilled that in all of us kids. "Shoot for the moon. You can do *anything* if you put your mind to it."

I was very young and idealistic. I thought the minute I moved to Los Angeles, I'd land my first acting job. The reality was different. After I relocated, it took about six months to get my first job. I moved, tried to find the right acting teacher, studied acting, got an agent. It took nearly half a year to get that first job.

I continued to study with different directors in Los Angeles, and at different workshops. One of the most influential directors for me was Milton Katselas. He's a wonderful teacher, and I still work with him.

When you first get a script, how do you analyze it? How do you begin developing your character?
Well, this may be wrong, but the first thing I do is look at the pages of dialogue. I look through the script for my character's name, to see how many times she speaks, how many pages she has. Before I even read it, I find out how involved she is in the script. That's the first thing I do. (Laughs.) The first thing you're *supposed* to do is read the script.

As an actress preparing for a part, I'll read the script, with nothing in my mind. No notions, no information. I'll read it fresh, like it's a book. Then I'll start from the "givens." I write down everything that is given in the script about the character. If I can make logical assumptions, I write those down.

After that, I write every question I have about the character. Ten questions. Then ten more. Those will spawn even more. I end up doing a full-blown biography. That helps me build her into a whole, believable person.

I do a lot of homework. And I believe in getting all the help I can get. I'll ask my acting teacher or my acting coach to read the script with me. They both know my work well, so they can direct me and be a mirror for me. It's hard to be your own mirror.

If the part is very difficult, I'll do a lot of research. Otherwise, it won't be necessary. For example, in the role I'm doing now for "Knots Landing," there wasn't a lot of research to be done. My character is under the witness protection program, so I got information about what it's like to live under those conditions. My character is married, has a child. I'm not married, don't have children. But that doesn't require research, it just requires imagination. In my workshop, I did a lot of improvisations on this character. That experimentation helped build up other sides to her.

When I did a guest appearance on the dramatic series, "Lou Grant," I played a reporter who was raped. That required a great deal of research. I talked to a lot of people on the Rape Crisis line, even answered phones, myself. Then I had to find out about being a reporter. I went to the newspaper, sat at a desk, talked to the people there. I wanted to observe the average day of a newspaper reporter.

So, if there's research I can do, it's great for me.

In "Roots" you played the part of Alex Haley's great grandmother as a young girl and an older woman, spanning sixty years. That was a monumental acting challenge. Can you tell us how you specifically researched and prepared for that part?

That was, without a doubt, the highlight of my acting career. "Roots" was history. It was based on real events, real people. Alex Haley spent twelve years of his life researching the character.

I had to do a great deal of background research for that role. I was a nineteen-year-old-girl in "Roots I," ended up eighty-something in "Roots II." Of course, I'd never lived in that time period, so I had to do a lot of reading. A great deal of my time was spent reading books about that time period. I needed to know how they dressed, how they behaved, how they lived.

I could also talk to Alex Haley about my character. He had a wealth of intriguing information. I learned that my character was the one that introduced colors to the slaves. She collected flowers, and from those she would make dye. She brought colors to the slaves and to the people in the big house. That's why she didn't have to go out in the field.

I read *Roots* over and over. I devoured all kinds of period books. I was constantly in the library, in rare book stores. I sat on the floor, looking at pictures, reading. And writing. I wrote it all down.

To get a personal portrait, Georg Stanford Brown and I went to visit old folks' homes. These folks were in the country and poor. I sat with my tape recorder and talked to some wonderful elderly women. They couldn't believe we came to visit, just to hear their stories.

I was fascinated, listening to them talk about growing up. And I watched like a hawk their mannerisms, their facial expressions. I watched how they got up out of a chair. How they sat down into a chair. It was pure gold. It was a real pleasure to do that kind of investigation.

How did you physically prepare for the aging process you portrayed in "Roots?"

That was very demanding. I had to make sure my posture changed as I grew older and that every part of me was real. There are certain things that happen to the body that haven't happened to mine. When you're working on your feet all day, your feet hurt. Maybe you get bunions. So I'd try things like putting rocks in my shoe. That way I'd never forget I had a bunion. It made me walk differently.

When you're older, your eyes change. I had to remember, am I nearsighted or farsighted? How does she look at things? With her head back? Close? All those decisions had to be made.

I talked to my parents, listening and watching. I was playing their age. It was interesting, my mom came to the set, I had my old-age makeup on, and was supposed to be her age, but she looked like my daughter. Of course, in the days of slavery, a twenty-year-old didn't look like a teenager. They'd look twice their age. It was a hard life, working from sunup to sundown.

I took a picture of my character to class, where we did a "picture exercise." I gave the photograph to the director, Milton Katselas. Then I'd strike the exact pose in the picture, getting into the physical and psychological state in the picture . . . as he watched. The feeling that emerges, is quite wonderful. You learn. You discover.

Is that the process you go through for all your roles?

It depends on the role. Sometime that physicalization doesn't happen until later on. If I can find out the character's physical traits, it helps a lot. I try to find her mannerisms, picture how she moves. What animal does she remind me of in her movements? Is it like a cat? A dog? Another animal? That is a real good key. That—and finding the sense of humor—are keys to characterization.

Sometimes I make conscious choices about how the character sounds. If she's a different character, like an aging slave, I have to find a placement in my voice, in my range, that will fit. But, if she's similar to me, then I use my own voice. I can use a lower register if I'm playing a secure woman, or put it up if she's less secure or frantic. It all depends on the complexity of the character.

Do you have input into the physical look and wardrobe of your characters? How do you work with costumers on "Knots Landing"?
Costumers can be very helpful. They see the character in a different way and provide a fresh perspective. They'll say, "I see this girl like this. . . ." And you realize, "Hmm. I missed that, but that's excellent." When that happens, it's ideal.

Sometimes costumers don't have an idea, or don't have the right tastes. In that case, I've got to be honest with them. I'm the one who'll suffer on screen. I know my body, my colors, and what looks good or not. It's my job to say, "This is not flattering to me."

The costumers on my show, "Knots Landing," have been doing the series for a decade. They know exactly what they're doing. The costumer is so full of ideas and taste, I'd let her dress me for anything.

You joined the regular cast of "Knots Landing" after nine years of successful series production. How did it feel coming into the show as a newcomer?

Lynne Moody as Patricia Williams in the CBS-TV dramatic series, "Knot's Landing." (Courtesy of Lorimar Telepictures)

That's an interesting question. There are positive and negative elements about joining a show that's been in progress for so long. One of the good things is the lack of pressure concerning ratings. When you're on a new show there's always pressure to get better ratings. The network may pick you up for six shows, or thirteen, or cancel you. So you can never relax. There's never that kind of tension in a long-running series.

Another good thing is that everything runs smoothly. Everyone knows their job very well, from executives to the crew. It's a well-oiled machine. And the actors are trusted, because whatever they've been doing, it's been successful.

One of the difficult aspects of joining the cast was finding a lack of enthusiasm. When I first came to the set, I was so excited. As soon as I'd come to work, I'd ask, "Did you see last night's show?" Everybody said, "Nope, missed it." They've been doing the show so long, it wasn't a novelty any more. They've lost that enthusiasm. But, it doesn't mean they've lost any ability to deliver. They sure produce effectively when the camera light goes on.

What changes have you noticed in your own acting styles over the years?
I used to be a selfish or "solo" actor. Those are actors who don't listen to you. They're self-involved, self-centered. I was one, because I didn't trust the other actor. I'd have everything worked out in my head. After a while, I realized that's exactly the way it looked on screen—like I'm acting by myself.

Now, I like ensembleness, sharing in a scene. When that's accomplished, both people look great. I think my work is better than ever, and I credit it largely to my acting workshops. That's where I can try new things, and fail. I learn from my failures.

I had to cultivate and develop my talent. Even if someone is a "natural" actor, they've got to learn the intricacies of acting for the camera. Workshops helped me learn how to project my feelings in front of this strange looking camera. So even "natural" actors can grow, learn, and develop.

What are your thoughts about working with Method-trained actors?
Some Method actors can be indulgent and selfish. If they take a half hour to prepare while I'm waiting, forget it. That's too indulgent for me. We all have to prepare. We prepare whenever and wherever we can. But I don't like to have other people waiting, doing nothing, while I prepare. That, to me, is indulgent.

One the other hand, Method actors are coming from a very honest base. I took several classes using that technique, so I know their emotional integrity. If they don't *feel* it, they won't say it. They won't open their mouths until they feel it, which can be a long time. It draws them right into the moment, it's real "moment-to-moment" acting.

What problems do you find working with directors in television and film?

I think television and film directors should know more about actors. They should go to workshops, see how actors work. They should be exposed to the process of acting. Unfortunately, many just seem to be experienced in the editing room and then get a shot at directing. Directors should understand the craft of acting.

They should also understand the terminology of acting. One of their infamous lines is, "more energy." What does that mean? They don't know, themselves. Directors need to know how to relate what they want. They need to know how to communicate with actors.

I would also love to work with a director who allows improvisation. Actors can discover a lot about who they are through improvisations. I've never worked with a director who permits it on the set. I've heard they exist, and I'm looking forward to that experience.

An actor and director want the same results—a good scene. If the director is inflexible, I may have to throw my preparation out the window. If there is some flexibility, I can usually find a way to incorporate my ideas into the director's. Unfortunately, I've done that and have been devastated by the results. (Laughs.) I've also done it my *own* way and been devastated by the outcome.

How would you compare the acting environments in stage, film, television?

Stage is the only place where actors have all the control. There's much more time to develop characters in rehearsal and during the run of the play. On stage, you don't have to rely on the editor, the cameraman, the soundman. What you see is what you get. And that's wonderful. It's instant gratification.

In television and film, you're at the mercy of everybody. When you're in front of a camera, you hope you're in focus. Then you hope you're lit properly. Then you hope the sound man is doing his job. And finally you hope the editor doesn't cut you out of the scene.

When you're shooting a scene, how do you deal with the pragmatics of repetitive takes and out-of-sequence shooting?

I like to get the scene right, won't settle for less. I'd insist on twenty takes if that's necessary. But, if I *have* gotten it, I don't want to shoot more takes for no reason. That's one of the problems I have with acting for commercials. They'll repeat scenes all day long, regardless. That could drive me crazy.

Shooting scenes out of sequence is a demand of the craft. I'll try to prepare with the help of my class. I'll do scenes that precede the one's we're shooting. Otherwise it's difficult to know where I am.

Let's say I'm doing two scenes. One requires me to be hysterical, the other requires me to be calm; the kind of calm that follows hysteria. I need to build to that moment, otherwise I can't do that hysteria, don't know what that calm is like, don't know how it feels to be drained. I need to experience that kind of calm, or that raspiness in the voice because I've been yelling. If I haven't done that hysterical scene, I might not build to a realistic performance.

Do you prefer playing close to type or different types of characters?
I don't have the luxury of doing a lot of picking and choosing, because there aren't a lot of parts for a woman, no less a *minority* woman. But the main thing that attracts me to a role is the honesty or believability of the character. I particularly like a character who is not a victim. Too often women are made to play victims. The role is appealing to me if the character's in a real situation, not because she's black, not because she's a woman, but because she's a human being.

I like to play parts very far away from me. When I played Alex Haley's great grandmother, it was demanding and very rewarding. The further away I get from my own type, the freer I am as an actress. It's wonderful.

I consider myself a dramatic actress, but I do get cast in a lot of comedies. The truth is, I have more fun doing comedy roles. Even if the part isn't great, I've been working on the lighter side all day. In drama, I'm drained by the end of the day. Dealing with emotional material is exhausting.

Has minority casting been a problem for you?
I know you haven't asked me directly about being a black actress, and I appreciate that. It reflects the way black actresses should be perceived, as actors capable of playing any roles. That's the reason I like the way they deal with my character on "Knots Landing." They don't explain I'm black, they don't comment on it or defend it. That's progress. It should have been this way a long time ago.

When I did "Soap," I was part of an interracial couple. I don't know if we were the first on television, but we were definitely the only one at that time. Interestingly, the producer said they never received any hate mail, except for one post card from Wisconsin. It said something about, "You may kiss niggers out there, but we don't do that up here."

That was the only negative reaction from the American public. Incredible. So the public is ready for it.

How do you feel about the audition process?
Company executives want to see what they're getting. But the audition process is often misleading. Some actors read well, then are terrible in production. They've taken classes to learn how to read. When they get the role, they don't know what to do with it. Other actors *don't* read well, but are great on the set.

Milton Katselas believed a good way to find out about the quality of actors is to talk with them. You can discover a person's quality, personality, feelings. I find that a rational way to cast. In fact, that's how I got my role in "Knot's Landing."

After starring in "Roots," you were unable to find work for a year. How could that have happened? What was your reaction?
"Roots" was a smash hit, a celebrated event, and yet *most* of that cast couldn't find work again. Even the star, Lavar Burton, who played Kunta Kinte, today has had a struggle. That's what got me angry. After being in the most popular show on television, to make that kind of history, you'd think we would have at least one job in the offing. And nothing happened. Not so much as one interview. I was very hurt. It pained me deeply, inside.

I had a lot of therapy to help me through that. I realized that jobs were not happening the way I dreamed they should, so I had to take some control over my life. I thought, "You're not making a great deal of money, you can't produce your own show, can't hire yourself." So, I decided to go to law school. I'd become a lawyer, because knowledge of the law is invaluable, and with a law degree I'd be able to fight back. I'd be able to have a say about things that happen to me.

I was accepted into law school in Seattle and was in the process of leaving Los Angeles. I told my agents. I told my acting teacher. I had a person looking for an apartment. I had price tags on all my furniture. That's when my agent sent me to one interview, for "Soap."

I took my audition to class, and Milton Katselas said, "What we need to work on is not your audition, but your attitude. You act like you're not going to get it." And I said, "Well, why should this be any different?" (Laughs.) To make a long story short, I got it, and put law school off for another lifetime.

How can an aspiring actor prepare for the insecurities of the profession? What would you advise them?

I would suggest very strongly you think about another line of work. You'll always hear about successful actors like Eddie Murphy or Jessica Lange, and how much money they make. You've got to realize you're seeing less than one percent of the actors out there.

When you come to Los Angeles, everybody that serves you food, that pumps your gas, is an actor. That's the ninety-nine percent you don't see. That's the incredible pool of talented people that are unemployed at any given time. And remember that the average annual salary of an actor is less than five thousand dollars, below poverty level.

I advise you to go to school, get an education, prepare for an alternative. That way you'll have some control. You'll have a say over what happens to you.

What advice would you give to new actors about training and working in the craft?
The most important thing is determination. I believe that if you hang in there long enough, sooner or later, you'll get a shot. Just make sure you're ready when that time comes. Develop yourself to the fullest, train yourself to capacity. Study hard. Take every lesson that's going to make you grow. Dance, acting, voice, dialect, accents, anything.

Taking class is very important, because an actor is like a dancer. If you stop working out, if you stop stretching, you'll stagnate. You won't grow. Commit to some kind of training.

You've got to study, work hard, and persevere. There are a lot of ingredients, but no single formula for success in this business. It's like going to the moon, if you mix things correctly, the rocket will go to the moon. I wish there were a set formula, but everybody has to find their own.

What is it about your own personality that allowed you to beat the odds and be a successful working actor?
Well, that's a really nice way to put it. I don't know if it's anything about my personality, or if it has to do with luck. I'm not going to fool myself and tell you that it doesn't have something to do with how I look. (Laughs.) I chose an attractive body, and I live in Hollywood where the physical is as important as anything else.

Luck is also important. Jimmy Stewart was asked that same question in an interview. For him it was a lot of luck. Things that are not in our control—like being in the right place at the right time—are vitally important.

I've worked very hard on my craft. When the ball came, it may have looked like I was fumbling, but I did get it past the goal line. I was ready. I trained hard, worked hard, and persevered.

I also learned how to adapt my life to the demands of acting. When I first started out, acting was my life. Everything I did was about acting. I wasn't happy until I had a job. When the job ended I was unhappy until I got my next job. So, I didn't really live until I started working.

Years later, I realized, "Wait a minute. I'm getting cheated. Life's going on, all these things are out there, and I'm singularly focused on acting." I was missing out on a lot. I wanted to change my attitude, didn't want to be cheated anymore.

Now when I get a job, it interrupts my life—and I have a life to come back to. It's important to laugh, to have friends, to share. It's also important to have other passions, besides acting.

Actors have to be committed. They have to work hard. But that doesn't mean they have to cheat themselves. Life's simply too short and precious for that.

Chapter **6** .

Harry Hamlin

Harry Hamlin discusses his concerns about episodic television, and explains why he took the role in "L.A. Law." He talks about film as a medium in which accidents create spontaneity and stresses instinct, imagination, and improvisation as the core of acting. He also discusses the problem of being typecast as a leading man, despite extensive classical training in ensemble acting. Harry Hamlin was interviewed in his trailer/dressing room during a break in shooting NBC-TV's drama series, "L.A. Law."

. . . .

Career Profile Harry Hamlin studied acting at University of California–Berkeley, Yale University, and with Lee Strasberg in a master class. He had several years of advanced training in the classics with American Conservatory Theatre, culminating in a leading role in *Equus* (1978). He was awarded a Fulbright Scholarship to study Shakespeare in London and played Hamlet in Princeton University's McCarter Theatre Company (1982).

Mr. Hamlin appeared in the film, *Clash of the Titans,* with Laurence Olivier, Claire Bloom, Maggie Smith, and Ursula Andress. His other films include *Movie Movie, Making Love, Blue Skies Again,* and *King of the Mountain.*

His specialty is long-form drama. He's starred in miniseries such as "Favorite Son," "Studs Lonigan," "Space," and "Master of the Game," and in television films, including "Laguna Heat" with Jason Robards and Rip Torn.

Producer Steven Bochco cast him in "L.A. Law" (1986), with the promise that the series would remain dramatically strong and that his character would have an opportunity to grow. Since that time he's starred as attorney Michael Kuzak in the award winning NBC-TV series.

* * * * * * *

Tell us about your formal training in acting.
I went through the entire U.C. Berkeley program in two years, transferred to Yale, and went through their drama program for the remaining two years of undergraduate training. I went through three years of classical training at the American Conservatory Theatre and became a member of the company.

I also studied with Lee Strasberg for a semester in New York, and with Milton Katselas for several years in Los Angeles. Most young people get started on their careers early, but I opted to study and learn my craft. I wasn't paid for acting until I was twenty-six years old.

When you get a new script, how do you break it down?
Each script is different, and each script requires a different process. My technique is a synthesis of all the different approaches I've learned. There's something a little sacred about discussing the process, because it's such a personal synthesis of those ideas. But I'll try to explain it.

My main concern is to find out what the character wants. All of his needs have to be fulfilled throughout the story, scene to scene. If I can figure out what he wants, I'll know where to go with character progression. I'll know if he wants to make somebody happy, get rich, get even. That basic "want" will help me determine how I approach the script.

There are usually enough clues in the script to give me a building block. There's a basic structure to figure it out. Then it's just building the character.

How do you go about building and physicalizing the role?
There are lots of techniques I use. The most common is an exercise that draws on imagination and observation of animals. I'll think of an animal that represents the character in some way. He's like a leopard, a lion, a bear. I can go to the zoo and watch that animal. I can develop a gait, a rhythm, or timing the animal has. And that's very useful. It's never failed me.

In *Movie Movie,* I played a rough and tumble fighter, who talked with a rough Chicago accent. I found the animal imagery and also researched the dialect. I put together tapes of Chicago accents from different ethnic groups and incorporated that into the role. To me, research is a great deal of fun.

In addition to research, you like to rely on instinct in your acting?
No matter how much research is done, no matter how much technique is used to flesh out a role, it's instinct that I rely on more than anything else. It's that first instinctive moment that comes in. That moment is like dancing with the muse, when all of a sudden (snaps fingers), you know what it's going to be.

Instinct can be muddied by method or technique. Some actors think their technique works, but it's halting their instinctual process. Actors must trust instinct or they're dead in the water.

What is your impression of Method actors?
Ideally, Method actors rely on instinct and use method techniques only when they're in trouble. Unfortunately, most of them rely too heavily on the Method. As a result, their performances are flat and one-dimensional. They're screening their own instinctual process by applying that external process.

It's a problem if they trust the Method more than their own core, their own connection to the muse. Much of that has to do with inspiration. It's hard for me to define, but without inspiration, no matter how much technique they have, actors will go flat in performance.

You preferred long-form drama to episodic television. What aspects of the "L.A. Law" pilot script appealed to you?
Steven Bochco gave me the pilot script and convinced me that this project would be different. Here was a very interesting piece of work that would allow me to stretch in a lot of subtle, dimensional ways. I got excited by the potential of exploring this one character over a long period of time.

The man I'm playing is in the process of discovering himself and making decisions about being upwardly mobile while still observing justice and honor in his profession.

In television, rehearsal time is limited. How do you deal with that in a weekly dramatic series like "L.A. Law"?
I prefer *not* to rehearse. I prefer to do the homework at home and to be spontaneous on the set. But, then again, I've already found the character. I've done the show for several years and know the character.

Harry Hamlin as Michael Kuzak in the NBC-TV dramatic series, "L.A. Law" (Courtesy of Twentieth Century Fox Film Corporation)

For a play, the rehearsal process is different. It's all about finding the character. Discovering interrelationships with other characters. Physicalizing the role within the set. Blocking it out. All that takes time. On stage, you know you'll be there every night, same time and place, to create that character. It's like doing a Pavlovian exercise.

What is your ideal relationship with the director? Should the director help you find the character?
In the last twenty years, directors have taken on an entirely different attitude. I've worked with some old timers who were great. George Cukor was a director who knew and visualized exactly what he wanted. He'd take the actor and approach him like moldable clay. He'd make the characterization, create the performance, and get it out of the actor.

That is not in vogue today. I think Method acting schools have diminished that approach. They want the actor to be the creator, the inventor of the performance. I have the feeling that performances may have suffered across the board over the last twenty years.

Certainly, the great actors are able to come out and be their own auteurs. They create the role and plug into the movie and make it great. But, in the old days, directors were able to get inspirational performances out of actors. They were able to inspire and motivate actors.

The ideal for me is a director who is not afraid to tell me his vision, not afraid to give me direction. Also one who is confident enough to be hands-off, to allow me to do my own thing if my instincts are right.

For the most part, I've had good relationships with directors. But I've worked with a few I didn't get along with. In those circumstances, I had to bite my lip, try to compromise, and go through with it.

Do you think improvisation is important?
I studied improvisation for two years, and think it's the most valuable study time that I had. I came away from that experience with a sense of confidence about my ability to work on the spur of the moment. I trust my ability to work well with other actors—without a structure, without having a script. It's a matter of give and take, particularly if there are three or four actors on a stage improvising a scene. In order for it to work, you've got to know when to give and take.

There are not too many directors who allow improvisation, which is a pity. I think improvisation is the kernel, the heart of what we do as actors. If we could do an entire picture based upon improvisation, and do it well, that would be the moment of real artistry. We would become artists because we were creating something spontaneously and in the moment.

Having worked on stage and film, what comparisons can you draw about the different acting environments?
In film, accidents happen. I find that a unique attribute of acting on film. On stage, one tries to avoid accidents at all costs. But on film, they create wonderful moments.

The best film acting involves an atmosphere in which accidents occur. It captures those things that are not rationally thought through. They're spontaneous moments, true to life. In real life we react without knowing how we're going to respond in advance. That's what can be captured on film.

I get as much satisfaction from working on stage as I do on film. But they are different kinds of experiences. Stage is more structured. It requires conditioning for the performance. I also can tap into artistic techniques I've spent eight years trying to learn.

On television and film, I don't use but a tenth of those techniques. But I do find television, film, and stage acting equally rewarding. When I'm on a roll in "L.A. Law," I get as much satisfaction out of that scene as if I were doing a great stage performance.

When you're working in film, do you prefer doing many takes or fewer takes?
If I get it in the first take, there's really no point in doing another one. But, directors like to have cover takes, just in case the first one has

problems. I have worked with directors who shoot fifteen or twenty takes just because they want to. Even if I had it in the first take, they want to see what happens in the fifteenth take.

That can get annoying, unless the director is willing to work with me. If I can say, "I'd like to try this in an entirely different way," the takes would be worth it. Instead of making the scene blue, I'm going to make it red. And if the director's willing to go along with that, I don't mind doing as many takes as I can. Of course, that presents a difficult problem for the editor in trying to construct a performance. Because there are so many different ways of doing it.

In "L.A. Law" you shot two different character portrayals of the same episode. Can you tell us about that experiment?
We did a show that was an existential crisis for my character. The director, Rick Wallace, let me shoot the show two different ways. We constructed the entire performance on two levels. We'd shoot it on the first level until it worked. Then we went back and did it again for the second level. I played it with a different twist.

They ended up using the first one. And the second one ended up in the can. An entire performance is out there on film that could have put an entirely different meaning to the whole show. It's in the can somewhere. Never seen.

How do you deal with the pragmatics of out-of-sequence shooting and rehearsals in film?
Out-of-sequence shooting is a fact of life. It's part of the preparation that you have to do. You've got to create the film sequences in your head, and all the moments and the beats that lead up to each scene in sequence. Then you'll know where you are at every moment.

It's not a matter of preparing how you're going to say the lines, because that's the kiss of death. It's a matter of learning a sequence of emotional events. So shooting out of sequence is not a problem for me. In fact, it sometimes creates interesting accidents.

That's why I don't like to rehearse in film. If you do, you're liable to fall into a groove, like a broken record. Every time it will be the same. If you don't rehearse, everything will seem as if it's happening for the first time. The words come out of your mouth for the first time. And who knows what's going to happen? Who knows how the other actor's going to react? Everything stays fresh.

Your training prepared you for character parts and repertory acting, yet you play leading men. Can you talk about that situation?

It's somewhat of a conundrum for me. I'm thought of as a leading man. But those parts don't usually require a lot of artistic challenge. I'm not called on to use the training and techniques that I know. For me, the most interesting roles are the ones that are physically different from me.

I was trained as a character actor, not as a leading man. All my training focused on repertory acting, ensemble acting, the ability to move from one role to another. Different people. Multifaceted parts. But when I moved into the professional world, the parts offered were much more limited. Producers and directors remember what you did last. It's a significant problem.

With the success of "L.A. Law," you should be in a good position to receive new projects. What do you look for in considering a new part?
With that kind of exposure, we do get a lot of attention and a lot of scripts submitted. But, that doesn't mean the scripts are any better. I look for three-dimensional characters in a script. Even if a script is good, there is no guarantee it will be produced. If the part's good, there may not be enough money to produce it. If there's money to produce it, the part may be weak. It's hard to find all the variables neatly fitting together.

How do you handle auditions?
I enjoy auditioning. A lot of times, I'll insist on auditioning, especially if I'm trying a role that's a departure. I'll tell the director, "I may have seemed like this in the last movie, but just watch. I'll do something that's very different." And I hope that's what the director is looking for in the character.

For those of us with classic training, auditions provide another chance to use those tools that we've gathered.

Before signing a long-term contract for your series, you worked from project to project. How do you deal with that kind of insecurity?
Actually, I prefer working that way. I prefer to go from project to project, like little islands in a big sea. I like the excitement of finding something and not knowing where I'm going to be for the next few months.

One of the reasons why I was reluctant to sign a long-term contract for "L.A. Law" was exactly that. It wasn't attractive for me to be doing the same thing year after year. It still isn't. I would prefer going from project to project. I like to be a free agent.

Sure, there's instability. But you can live your life not affected by that instability. The overhead costs of your personal life can be reduced to live that way. My wife and I have maintained that minimalistic style. When "L.A. Law" ends, we'll be able to live the same way we lived before. Maybe we'll work, maybe we won't, but we're not going to be upset about it. It's not going to really affect our lives.

The reality is, the down time is psychologically difficult. You can be swimming around, not knowing where you are. But if an actor has been trained, and there's a solid foundation to your work, there will always be a job somewhere. You may not know where it's going to be, or what the calibre will be, but there will be a job.

What advice would you give to students of acting?
First and foremost, become a doctor. (Laughs.) I believe that becoming successful is probably ten percent talent, ten percent training—and eighty percent luck. If you're in the right place at the right time, you'll do it.

There are thousands and thousands of extremely talented actors who can't get their foot in the door, for one reason or another. It's because they weren't introduced to somebody, or weren't in the right town, or weren't sitting next to someone who's directing a new film. It's a matter of where the dice fall. There are so many people who are very talented who don't ever get a chance.

If you're really serious about becoming successful as an actor, perseverance is probably *the* most important thing. Don't ever say, "I can't do this." Because that will be the end.

Once you've been trained, keep in shape. Keep yourself ready. Keep working on your craft. In the event the phone rings and somebody says, "I've got a job for you," you've got to be in shape.

What do you think there is about your personality that enabled you to beat the odds?
It was eighty percent luck. Being in the right place at the right time. I certainly patterned my life in order to get there. I worked hard, trained hard. I was ready when the shot came.

I also learned a very valuable lesson from my father—trust your own instinct. He championed that idea, to trust the little voice inside. And I advise everyone to keep attuned to their own inner voices. If you can tap into that, there's no end to what you can do. Whether you're trying to be an actor, writer, businessman, stock broker, you name it. Do what's best for you.

And never go against your own instinct.

Chapter 7 .

Leslie Charleson

Leslie Charleson talks about her early studies with Uta Hagen and Herbert Berghoff, and discusses the challenges in playing the same role for more than ten years. She analyzes the work environment in daytime drama, describes a sample production schedule for videotape drama, and compares the pragmatics of acting for daytime with primetime television. Leslie Charleson was interviewed in her dressing room at ABC Studios during a break in shooting the daytime drama series "General Hospital."

. . . .

Career Profile   Leslie Charleson is an actress who has worked regularly in daytime drama, as well as episodic television. She's played the part of Dr. Monica Quartermaine on the ABC-TV soap opera, "General Hospital" for over ten years, earning three Emmy nominations as Best Actress.

Ms. Charleson attended Bennet Junior College in Millbrook, New York, graduating with a degree in fine arts and an award for outstanding actress. In New York, she studied with Uta Hagen and Herbert Berghoff, and at the American Place Theatre.

In her first few months in New York, she was signed to play the part of Pam Corso in the ABC daytime drama, "A Time For Us." After six months, she left the show and appeared in numerous commercials, including ones for Pearl Drops Toothpaste and Polaroid Swinger Camera with Ali McGraw.

When daytime drama was still live, she landed a role as Iris Garrison in the CBS-TV's soap, "Love Is A Many Splendored Thing." She played the same character for nearly three years. At one time, her character was simultaneously blind, pregnant, and terminally ill. She recovered from everything and earned herself her first Emmy nomination for Best Actress in the process.

Ms. Charleson moved to Los Angeles, where she guest starred in a number of episodic television shows. Among them: "The Wild Wild West," "Marcus Welby, M.D.," "Adam 12," "Cannon," "Search," "The Rookies," "Emergency," "Medical Center," "Owen Marshall," "The Streets of San Francisco," "World of Mystery," "Ironside," "Kung Fu," "Caribe," "Medical Story," "Barnaby Jones," "The Rockford Files," "Happy Days," "Baa Baa Black Sheep," "Most Wanted," and others.

During this time, director Mike Nichols, remembering her from the Pearl Drops commercials, cast her in the feature, *Day Of The Dolphin*. She starred opposite George C. Scott in that film.

She joined the cast of "General Hospital" in 1977, playing the same role for more than a decade.

. . . .

Tell us about your background and training. You majored in theatre and worked in summer stock?
I wanted to be an actor since I was a little girl, and couldn't think of anything else to do with my life. So I majored in theater at Bennet Junior College in upstate New York. It no longer exists, but I chose it because it had a good fine arts program, it was two hours outside New York City, and I could get right into my major. I just wanted to get to New York. Fast. I chose the least painful way.

When I came to New York, my parents said they'd support me for three months of acting classes. Then I'd have to get a job in something other than my "chosen profession." I studied with Uta Hagen and Herbert Berghoff (H-B Studios); I also studied at the American Place Theater. I was lucky, got a job three months after I moved to New York and haven't had many employment gaps since.

During college, I did summer stock. Albert Brooks and Rob Reiner and I had a wonderful summer at Cape Cod doing summer stock. We laughed so much, I thought I'd get lockjaw. I learned a lot about the intricacies of theatre in summer stock. (Laughs.) Want me to build a flat for you? Or size it? Don't ask me to do costumes, I'm a bust at sewing.

When I studied at H-B Studios I lived with three other gals. One went to a school that didn't let you accept a regular job for two years. That wasn't for me. I felt it defeated the purpose of acting school. At the H-B Studios, I worked with real professionals. I was in dance class with people who were doing Broadway shows. They were professionals who were "tuning up" and keeping at it.

There is nothing better than working with people who are better than you. It makes you work. That's what I appreciated about the school. I had a chance to work with those who were much better than I was. That's where the challenge and excitement came in.

What types of roles did you work on during your studies with Uta Hagen and Herbert Berghoff?
Everything. That's the luxury of taking acting classes. You can do things you'd never be cast for—Medea, Lady MacBeth, everything that stretches and challenges you.

Contrast that with all the typecasting of nighttime television roles. I always guest star as the "good" wife or the "nice" girlfriend because I was blond and pleasant looking. I was always the "good" person. I didn't get any of the nasty roles because I didn't dye my hair brown. If they'd only just be creative and think about casting. Class is where you can go and take risks without worry.

You've done the same part, Monica, in "General Hospital" for over a decade. How do you feel about that as an actress?
I celebrated the tenth anniversary in 1987. Initially, the contract was only for two years, but it worked out differently. (Laughs.) Time sure goes by when you're having fun. I'm lucky to have a good group of people to work with. I wouldn't be there without them. It's been challenging and fun.

The fact that I've been here so long belies a reality about "stability" in daytime drama. I'm afraid stability is an illusion. Contracts are meaningless. You can't say, "I'm leaving." They *tell* you when you're leaving.

Someone else can play Monica, I have to face up to that reality. The character of Monica is what I make her, but I didn't create her. Someone else created her almost two years before I came in. We're not indispensable. The minute actors begin to believe their fan mail, that's the beginning of the end.

We've had a character named Heather. We've had *five* Heathers since I've been here. There are many character turnovers. The audience may get cranky at the turnover, but they'll still watch. The soaps are on every day in millions of households.

Nonetheless, I love the part of Monica, and enjoyed performing it all these years.

When you get a script, you're at some advantage because you don't have to start digging into a new character. How do you approach the script?
Last night I received two scripts for next week's shows. I read them to see how many scenes I have, to get an idea of how much work I'm in for. For next week's shows, I don't start learning lines. I can learn lines very quickly.

A long time ago I learned—doing *live* soaps—that it's very easy to get confused about scripts that seem to duplicate a lot of lines. I remember walking into one scene, thinking, "I'm doing yesterday's show." (Laughs.) I looked at the woman opposite me and thought, "What day am I saying?" I've said these words before, to other people, and didn't know whether I was performing for the first time or repeating dialogue. It scared me to death. The problem was that I prelearned things, got too many scripts confused.

Now I prefer to learn the dialogue the night before we shoot. I don't like to get too familiar with the material because there's too much time to find flaws in the writing and continuity.

What can you do if the dialogue or character action doesn't make sense in the script?
Try to make it work. It's the only way to get on with the production. I have to say the lines, or we'll be here for months. That's the name of the game. If I'm really uncomfortable with something, I'll try to work it out with the other actors and director. Otherwise, I'll try and make it work, even if I don't know what I'm talking about.

There are many times where the writing is weak and I don't think anyone can read it credibly. Still, we have to say the lines, it's our job as actors. Once it finally becomes acceptable, it becomes our own. And there's a better than fifty percent chance we might actually get the show done.

When you analyze the script in terms of your own character, what do you look for to help physicalize the role?
I read the whole script, and pull information about my scenes. I like to know if an actor in another scene may be talking about me. It gives

a hint of where I am and what frame of mind I'm in. If someone says, "Oh, Monica is exhausted," then I don't want to be starting out peppy in that scene. If everyone is talking about how tired I am, it gives me an idea, it gives me a clue about my physical and mental state.

After I read the script and pull my scenes, I try to envision the other actor. If it's the person playing my husband, I know him so well now after ten years that I can almost hear what he's going to say, or what he's not going to say. It makes it a lot easier.

I try to imagine what the scene will be like, including picturing the sets and how I'll be moving. That takes care of a lot of initial concerns about location and wondering where I'll be standing, and with whom. I learn the actual blocking from the director.

I learn the role to a point, but it's not until I hear the other actor and get the stage direction that it all comes into focus. One fits into another. You associate words with moves, with people, then another actor's going to come in and say something slightly different. Words aren't, fortunately, engraved in stone. And that's where we start. We spend the rest of the time "running lines," operative phrase. There's no help from the director, there's simply no time.

You know your character very well, obviously. What happens if you get a script that doesn't quite work for you. Your character wouldn't do it, the situation isn't quite right.
(Laughs.) We're talking about today, for instance. Sometimes, I think, in all fairness to the writers that have a big task with a big cast, they don't know the characters. There are times I feel they've hired demented chimpanzees to write the show, or at least people who've never seen the show. That makes it difficult for us, because it's harder for us to work through a bad script.

It's comparatively easy to do Harold Pinter, Edward Albee, or Tennessee Williams. That's easy. The words are there and you don't mess around with them. In our case, you spend most of the day trying to get the words to where you can say them without feeling like you have to wear a paper bag over your head; or that you'll have to change your name and move.

There are certain character traits that should hold true in all present and future scripts. If they're violated, it hurts the characterization badly. For instance, there was a character on the show for years and years. It was established that he was an alcoholic, and of course, every script mentioned that he wasn't drinking.

Then we shifted writers and the first script comes in and the character is making martinis and drinking them. That's when you barnstorm the office and scream, "Hold it!" It's an affront because the actors have worked so hard to establish the characters.

You hope that writers will see what you've brought to the series and work with you. It's frustrating, and a lot more work, when you're pitted *against* each other. I do my homework, and hopefully writers will all do theirs. Learn the characters. Only then can our interests mesh.

How do you feel about working with someone who has been trained in the Method?
If that person works *indulgently,* he'll get over that on the soap, real quick! There's no *time* for that indulgence on the set. On daytime you don't put on false beards, wigs, mustaches, you don't age physically. The characters are who you really are. You have to speak the words given, and incorporate them into your own character.

Anyone who doesn't fit the role instinctively is in big trouble. Actors have to come on the set, ready to go, no matter what. Homework has to be done at home.

You've worked in daytime drama as well as primetime television and film. How would you compare the pragmatics of those acting environments?
Working on a soap opera is extremely hard work. Many film actors who do guest shots simply don't understand how we can keep it up. In one day we shoot a one-hour show. Compare that to the seven or eight days required for primetime. In one day, we shoot ninety or a hundred pages of script—as opposed to five or six pages shot in a good day of primetime film production. We work every week, and we never have hiatus.

It takes a certain kind of actor who can function under that kind of pressure. That's why it has to stay interesting and rewarding. Otherwise, it's too much of a sacrifice. There's no time for other things.

Would you prefer working in daytime or primetime television?
A lot of daytime actors think nighttime is "big time." Then they try it and come right back to soaps again. I did nighttime, then came back to a soap. (Laughs.) Nighttime is nothing more than daytime but with a little more lip gloss and shoulder pads. It doesn't have the energy we do.

I'd much rather do soap opera than do primetime television. To me it beats the hell out of sitting in a trailer from 6:30 in the morning to

6:30 at night, only to have a stage manager bark, "Sorry, we can't get to you today. See you tomorrow." You'll have one line and they'll get to you in the evening when your brain is marshmallow. That's no fun.

I think most actors will say the same thing. I would rather have forty pages of dialogue in daytime than enter in the second act of a play and say, "Aye, my sire." Waiting to come on, I'd be gone. My mouth wouldn't work, and my brain wouldn't function.

As a matter of fact, that happened when I was doing a scene with Rock Hudson. I sat in my trailer from 6:30 in the morning to 6:30 at night when they finally said, "we're ready for you." The crew was gathering up the cable, and I had my one line to say to Rock Hudson: "There's no doctor on board, sir." They called "ACTION!" I walked up to Rock Hudson and said, "There's no **B**octor on **D**oard, sir." (Laughs ironically.) I didn't know why everyone cracked up. I twisted the words, but it sounded fine to me. Sure, it sounded fine around two in the afternoon, then slipped into this dialectical mush.

That's not acting, that's an endurance test. There's really no thinking of a character, you just have to worry about your mouth working.

It's a very different experience in daytime soaps. I have forty-seven pages today, with a huge number of lines, and I manage to crank through the whole thing. It's very difficult, but I'd much rather deal with that overload than just one line of dialogue. I don't want to have just one line. For that reason, my hat's off to extras.

Film actors contend with out-of-sequence scenes, but daytime actors also face out-of-sequence shooting of episodes. How do you handle that?
It's one of the most frustrating things to be shooting out of sequence. I panic about doing something that is not relevant to this episode, or repeating something, or abandoning something that's needed. It's hard to keep track of it all.

For instance, we're doing show No. 113 and then today I'm shooting No. 131. Yesterday, I did show No. 115. What happens in between?! That's terrifying, because I need to make sure that I've gotten the information and won't do something silly.

In film, scenes are shot out of sequence, but you have the whole script. That's okay, because you can pick up the script and go back and see where you would have been. If you're doing the end of the film at the beginning, and the beginning of the film at the end, at least you know what the first, second, and third acts are. We *never* have a third act of the show.

Doesn't the director help you focus and clarify the contexts of the scenes?

Not at all, they don't have time. We have three directors who basically are there to direct the technical aspects of the show. They're about as informed as we are. We've been shooting so many remotes lately that we're terribly out of context. Some actors have flip-flopped scenes and shows, it's easy to do. We're all a bit confused.

Don't you have an outline of future episodes? A sense of story progression for the characters?

No, because there isn't any way to know what the audience will respond to. The audience response, ratings, and fan mail determines the show's direction. The network and producers can bump off a storyline just like that if they hear a negative response, or if they see that the ratings have dropped.

They have no compunction about switching gears with actors. It has no bearing on whether an actor is good or bad, it's just that the storyline isn't working out, so they're going to get rid of that character as fast as they can. That's why there are so many people on the soap.

They always have what's called a "front-burner." That's what they're exploiting at the moment, but in their back pocket is another storyline that is great for Wednesdays and Fridays. It's that sort of thing. It's very hard. I don't envy anyone that's out there in the production office at the moment.

What is a typical production schedule for you in daytime drama?

Let's look at today, for example. I've been here since 7:00 A.M. We hit the stage, and the director blocked the first thirteen scenes, without cameras. It's rather like : "Okay, Monica, on page thirteen you'll cross from the door to the center desk. Got it? At the end, you'll cross to the mantle, then cross back when you say this dialogue on page twenty. Then you'll leave out the door."

It's that quick, and I'm searching desperately for a pencil to write all the blocking down. (Laughs.) Then I race downstairs, someone yanks me into hair and makeup, someone hurls me into wardrobe. At 8:15 A.M., we're flung on stage with the cameras, so that they can see what's going on. We flounder our way through a camera blocking, usually not very well, then we race downstairs, finish our makeup, hair, and wardrobe.

By 9:30 A.M. we're standing on stage, someone says, "Tape is rolling!" I have yet to say "good morning" to the actor I'm working with.

There isn't much time to rehearse. It's very difficult to have lots of scenes in a row and not have the opportunity to rehearse. I wish we could, but we can't.

What are some other production considerations in taping daytime shows?
There are some exciting moments in production, when you strive for the brilliant moment in a scene and hope that it was actually captured. There are times when you feel really good about a scene, a moment when you think, "This is great! I've gotten into it right! It's 'clicked' for me."

You feel especially good if the crew claps, because they don't care if you come back the next day. Directors will tell you it's good all the time. They want you to come back with your hair brushed and your lines learned.

Unfortunately, sometimes those magic moments are lost to posterity. All those brilliant moments might be lost because the camera was on the other person. Or little fine touches you were doing with your hands are lost because they were shooting around chin level, and no one will ever know. It certainly can take something away from you.

What type of role would you like to do next?
I'd like to do something that I can be very proud of. I'm open to anything that stretches me. It could be film, television, theatre, anything. As long as it's acting.

What would your advice be to new actors?
You've got to try things out, everywhere and anywhere. Do community theater, do summer stock. That weeds out the "men from the boys" and the actors from those who complain "Hey, this is work."

I get so many letters from people saying, "I want to be a soap opera star." It sends me right up the wall. There is no such animal. We're actors, all of us, and we work hard. Some aspiring actors look at our work and say, "I can do what you do, it's easy. You come in, sit down, talk, have a cup of coffee and danish. . . ." It's a compliment to us that we make it look so easy. It requires a lot of training, and it's a lot of pressure.

Daytime television is very hard to do, but it is one of the best training grounds for actors. There's no other place you can get paid for acting lessons. Yesterday I could have blown my performance and been cranky with myself, but today I have a chance to come back and have another shot at it. Different words, different time. I can pull my knickers up and get back in there.

There's no better training for acting than working in daytime soaps. You can't get any better experience, and you can't have anything else that approaches the semblance of "stability"—which there isn't. You do have a chance to work at your craft. And there's no way you can get anywhere if you don't keep doing it.

What about the insecurities of the profession?
I remember working with George C. Scott on the film, *The Day of the Dolphin*. He was sorely bemoaning the fate of actors. He was sitting on his boat, angry about the business, telling us all to get out of the industry. It's degrading , demeaning, sallow, and humiliating, nothing good about it. I finally turned to him and said, "tell me G.C., what are you doing in the business?" And he replied, "It's the only damn thing I know how to do."

It *does* frustrate you. We are all basically disturbed people. We are very fragile, with delicate psyches, and yet we expose ourselves. It's not like we can hand somebody a book and say, "I wrote this, read it"—it doesn't matter what we look like. Or "I painted this, view my painting"—it doesn't matter what we look like. The fact is, actors stand up there and let people judge us—you're too fat, too thin, too tall, too short. We put ourselves in the most vulnerable position, one we're terrified of being in.

It sounds like perseverance is a prerequisite for success in your eyes.
Perseverance and passion! I do resent people who are very casual about wanting to be an actor. Their attitude is so blasé, "I might be interested in acting, I'd like to be a star, so I'm thinking about it." That's when you better watch me, because I'll go for someone's throat. You have to have a *passion* for acting.

I protect what I do because I've worked hard. I've paid my dues. But you can never pay enough dues. You have to love the craft and really be sincere about it. It's not a casual commitment, you can't just shrug and say, "Well, what can I try today?" Acting is extremely demanding, it takes a lot of work.

What kind of training do you think best prepares people for those demands?
My advice is to get an education. Finish school, and do summer theater—that's a great experience. When you go to college, major in

theatre, because that's your only shot at reading important dramas from all ages—Classics, Romantics, Greek tragedies, Comedies, Shakespeare. Study them, know them. Read all the dramatic plays you can. That's your opportunity. All that homework is necessary for a basic background in acting.

You need to get that education so you know what you are talking about. Then study and work. Take classes, from improvisation, to scene study, to cold readings. Do it.

For people who aren't established, you have the luxury of doing anything and not having to take any risks. Nothing's involved. People won't say, "Tsk, tsk, look at Leslie Charleson, wasn't that a bomb." The more you do in *every* area, the better. Work backstage, in summer stock, be an apprentice. Watch, observe, read.

Do *anything* and *everything* to help yourself succeed.

Chapter **8** ·

Jack Gilford

Jack Gilford talks about the importance of imitation in acting and describes his transition from vaudeville to theatre to film. He analyzes characters played on Broadway and talks about the value of a committed director. He explains how he improvised the ending to the film, Cocoon, *and recalls the network's persistence in typecasting him in a series pilot. Jack Gilford was interviewed at a cafe in Washington, D.C., prior to a stage performance at the John F. Kennedy Center for the Performing Arts.*

· · · · · · ·

Career Profile Jack Gilford is an actor with wide experience in all aspects of performing, including vaudeville, nightclubs, films, television, and Broadway. He's a veteran comedian who has played comedy and drama with equal ease. He's earned Tony Awards for his comedy work on stage and an Academy Award nomination for his dramatic work in film.

Mr. Gilford worked independently to perfect his talent for mimicry. He performed in vaudeville and nightclubs, miming some of the period's most renowned personalities. His first role on Broadway was in *Meet The People* (1940). He won his first Tony nomination for his role in the musical comedy, *A Funny Thing Happened On The Way To The Forum,* with Zero Mostel, and recreated that role in the motion picture adaptation (1965). He appeared on Broadway in *No, No Nanette,* and *Cabaret,* which earned him another Tony nomination.

Among his stage achievements, Mr. Gilford was the first nonsinger to play a major speaking role in a Metropolitan Opera production, *Fledermaus* (1950). He performed in stage and television adaptations of *Once Upon A Mattress,* with Carol Burnett. He considers his characterization of the 87-year-old Jethro Crouch in *Sly Fox* (1976), with George C. Scott, a high point in his comedy career.

His dramatic roles have won critical acclaim. He created the role of Mr. Dussell in *The Diary Of Anne Frank,* Mr. Zitorsky in Paddy Chayefsky's *The Tenth Man,* and Bontsche Schweig in *The World Of Sholom Aleichem.* In 1987, he appeared on Broadway in *The Supporting Cast* with Hope Lange and Sandy Dennis.

On television, Mr. Gilford appeared in a great number of dramatic and comedy series and specials, including "The Defenders," "Of Thee I Sing," "The Carol Burnett Show," "Dean Martin," "Trapper John, M.D.," "All in the Family," "Rhoda," "Love Boat," "Taxi," and "Soap." He was seen regularly as Grandpa Hollycock in "Apple Pie," and Brooks Carmichael in NBC-TV's "The Duck Factory" (1985). He co-starred with Dom DeLuise in the television film, "Happy" (1986), and was featured in "The Very Special Jack Gilford Special" (1986) with Anita Gillette for CBS Cable. He also created a successful run of "Cracker Jack" commercials in 1986.

In motion pictures, Mr. Gilford has been equally active. He's appeared in dramatic and comedy films including *Enter Laughing, The Incident, Catch 22, They Might Be Giants, Who's Minding the Mint, Caveman, Cheaper to Keep Her,* and *Save the Tiger,* co-starring Jack Lemmon, which earned him an Academy Award nomination (1972). He most recently co-starred in *Cocoon* (1987), its sequel, *Cocoon II,* and *Arthur II.*

· · · · · · ·

Tell us about your background and the experiences that influenced you as an actor.
I didn't have any formal education in acting, but I've always had a keen imitative sense, and I always worked to sharpen it. Ever since I was a child, I worked on imitating people to get a laugh. In my first act in vaudeville, I imitated some of the biggest names of the day—Harry Langdon, Rudy Vallee, George Jessell, Laurel and Hardy, Charles Butterworth, Jimmy Walker, Al Jolson, John D. Rockefeller, Sr.

One of the biggest influences on me was Charlie Chaplin. When I was five years old, I went to see a Charlie Chaplin movie and it changed my whole life. It was a new picture about gypsies. He had his back to the camera, and I somehow knew exactly what he was thinking and feeling. He was a brilliant actor. I wanted to be just like him.

I'd go home and say to myself, I think I can do that. (Laughs.) Never mind I could never achieve it, but *trying* thrilled me. I'd watch people like Spencer Tracy, and I was enthralled. I knew what he was *thinking* on screen. That was magical! There was no choice, I had to be in this business.

Without consciously drawing on that every time I perform, I realize how important it is to observe the way people really behave, and to create it realistically.

The best teacher was hands-on experience, *doing* it. I was fortunate to be in several long-running theatrical hits. There's certainly a lot you can learn about acting from doing two-year runs. I did that in *Diary, Forum, Cabaret,* and many one-year runs, as well. You can really get to know a character in that length of time.

Your background was in vaudeville and nightclubs. How did you move into theatre?

The leap to the legitimate stage was a significant and qualitative one. I was helped enormously by director Garson Kanin, who saw me in a small, Left Wing review during the war, *Meet The People* (1940). He just got back from England, where he was directing government movies. I was still playing nightclubs when he asked me to be in one of his plays. The show wasn't that successful, but it gave me a chance to ''act.''

I guess Garson liked what he saw. While we were on tour, he adapted *Fledermaus* from all the other infamous adaptations of the opera. He was doing it for Rudolph Bing's first season at the Metropolitan Opera House (1950). The papers announced that Danny Kaye was going to play a starring, speaking, comedy role. I remember thinking how huge that place must be. I preferred small places, like nightclubs. To me, nightclubs seemed more like chamber music than symphony.

Then Garson phoned and asked if I would be interested in playing the part because Danny Kaye was unable to do it. He asked me to read the script. I laughed and said, ''No. First I'll accept the job, *then* I'll read the script.'' The script was wonderful, and it was truly a funny part. Up until then, the roles I played were summer stock type, not too funny. I'd attempt to make them funny, but it was like lifting a piano. You get it off the floor, it's not far enough. I was happy to accept the job, played the part at the Met, and eventually became better known as a comedian who could actually act on stage.

But after I worked the Met, I was suddenly blacklisted from performing. Our country entered the blacklisting era, and I was considered a person of "left" persuasion. During that period, a lot of creative people suffered. Two of them worked on a touching play—Arnold Pearl, the very fine writer, and Howard DeSilva, a fine actor and director. The play was called *The World of Sholom Aleichem*. It was actually three plays in one, dealing with the time of Yiddish writer, Sholom Aleichem.

In that play, I played a tragic character named Bontsche Schweig. Bontsche just sat and looked at the world and had a smashing, shattering line that ended the play. I wondered how they chose me for this deeply moving part. Maybe they saw me do pantomime in my own act. Maybe they saw something in my conscious or unconscious character. I don't know. But it was a wonderful match, and I fully understood the tragedy of this poor man.

How did you create the character of Bontsche in The World of Sholom Aleichem?
The script said it all, and I felt at one with the pathos. Bontsche was a tortured character who finally wound up in heaven. The angels made such a fuss when they heard Bontsche Schweig was coming to heaven. I came in with my feet wrapped in burlap, I looked haggard, with a nothing kind of beard. They sat me down, and Ruby Dee, the defending angel, set my case before Morris Carnovsky, who played God. The prosecuting angel started his case against me, but just couldn't continue.

God pointed to the heavens, "Bontsche, take what you want, you deserve it." I could have had anything. Anything. I took a long look at heaven, and then turned back, "In that case, if it's true—can I have a hot roll?"

The first two weeks of the show, I had to hold back tears when I did that line. I felt my character mustn't let the others see that I felt something deeply. It added an intangible strength to the final moment of the play. It was a shattering moment.

When you approach a character for the first time in a script, how do you get to know him? How do you begin to analyze the script?
It's all in the text of the play. I've learned it's what *other* characters say about you that's vitally important. You get an idea of who you are by reviewing what others in the script say about you. You can start from that perception and then learn more by saying the dialogue, feeling the words and thoughts. That develops your style and taste in acting.

How do you build the character from that point? What values and choices do you make?

Without direction—which I need and depend on—I have to rely on my own instincts. I "assume" the character and just feel the truth of what that character would do and say. I have no method, no guide, no measure. It really depends on my own taste.

That's a peculiar thing—taste in acting. Actors call it "selection." You select something and go for it. You could be wrong, you could be right. If you're wrong, you can correct it before opening night. If you're right, you stay with it.

An artist friend asked what the guide is for taste. There is no guide. It's purely personal. If you're lucky, bright, and good enough, you'll find the right values for that character.

How do you physicalize a character? Can you give some examples? Let's start with The Diary of Anne Frank.

I played the part of Mr. Dussell, a dentist in the play, *Diary of Anne Frank.* I read the script and asked, "What kind of a man is this? What should I know about him? Garson Kanin was the director, and he saw him as a pretty cranky character, with an undertone of comedy to his meanness. As a director, Garson wanted to get as many laughs into the dreary tone of the play as possible. Luckily, I picked some things to help create that kind of character, selecting eccentric physical character traits in rehearsal. I stayed within the context of the character written in the script, even though I deepened it.

At one point in that show, Susan Strasberg and Dan Levin were supposed to be on their first date in her room. The rooms were incredibly small, and they were angry at me for being in their way. Each of them slammed their bedroom door in anger, I reacted, we had a blackout, and a burst of laughter.

Garson wanted to know what I was doing to produce that laugh. Do you know what I did? A cat. We had a peculiar cat at home that would make mad dashes, running to the couch. It was that animal, cat-like movement I did. The slam here and the slam there was loud enough for me to stand quietly and react like a cat. It was an instinctive selection that worked well.

Sometimes, my choices are more obvious. I'll do some line or physical action, and I'll think, who was that? Where did it come from? I once did a character in *The Student Prince* who was an absolutely outrageous character. So I invented an outrageous way to say the lines. The way I said the lines reminded me of someone, but I couldn't place it. Then, finally I realized I was imitating Eddie Cantor. I was doing this line with great surprise, just like Eddie Cantor. At other times, I found myself imitating the physical style of Bert Lahr, another great comedian.

It's interesting to analyze because most of this is subconscious. Yet I remember making a deliberate choice about the way I would sing Eddie Cantor's "Whoopee" in another stage production. After great deliberation, I decided to sing it just like Eddie Cantor, because nobody was using his style anymore. Why shouldn't we use it? I gave myself that reason for doing it. The audience became reacquainted with Eddie Cantor's style, and my own character was more interesting in the process.

So the imitative skill works in helping to physicalize a character. I wouldn't stay in that exclusively, but it does help if it's appropriate.

What is your reaction to the Method as a technique for actors?
I haven't used the Method, but I had an early lesson from one of the great Method teachers, Sanford Meisner. We were in a farce together back in 1942. I played the part of a lawyer who shared an office with a guy who was into stolen goods. My character was a lawyer who never won a case. I had to make an entrance from the office doorway, and Sandy Meisner asked me where I was coming from. I said I'm coming from the doorway. (Laughs.) He looked exasperated!

Then I got it. That's what the Method is—knowing what happened to you before you came in. What happened to your mother? Your father? Is that crazy cousin really crazy? I always felt—even without the Method—that if a character had a peculiar name or traumatic experience, it would have an enduring influence on his life right now.

The Method can be very effective as a tool for some actors. Robert DeNiro studied the Method and is a magnificient actor. He was in *Mean Streets,* a film about Little Italy, directed by Martin Scorsese. When I watched that film, I forgot I was watching actors, who had to "assume" something before the cameras rolled. It seemed like Scorsese got these guys off the street, and they forgot the camera was capturing them. These guys *were* the part. So I have respect for the Method.

You've created characters for the theatre and adapted the roles for television and film. How do you translate your characters for film?
It's an interesting question. Having created roles for the stage, I had to revise my thinking about the characters for film adaptations. I play

much broader on the stage. It's a challenging situation. The first few days of rehearsal, the director has to tone me down—vocally, physically—especially if I just came from the stage. In film, there is the added opportunity to think, to make things more subtle, more intimate.

Jack Gilford as Phil Green in the motion picture, *Save the Tiger*. The role earned him an Academy Award nomination. (Courtesy of Richard Benson)

In theatre, without the time constraints of film, the director should have more time to work with you. Do you find that generally true? It depends on the director, and it's very frustrating not to have feedback. I didn't have much direction in *Fledermaus*. I constantly asked Garson Kanin for feedback, but he complained he was too busy. He was having enough trouble with those "so-called actors" in the opera—big stars like Richard Tucker, Richard Brownley. He left me alone and spent most of his time and attention on everyone else. Looking back, I guess he must have approved of what I was doing on stage.

The same thing happened when he directed me in *Diary of Anne Frank*. I'd come home and my wife would say, "How are you doing?" I'd shrug, "I don't know, Garson doesn't tell me." So I finally confronted him. He said he liked what I was doing. As a director, he did some innovative things for us. We started rehearsing on a hot August night. The design of the set was taped on the floor of the stage. He wanted the cast to wear three layers of clothing, so we'd move with the burden of the characters. That really helped. He said he'd help us if we got all tangled up. (Laughs.) And we sure did.

I love to work with a creative director. You can experiment in rehearsals and find the character through your artistic taste and choices. Of course, whatever you do on stage eventually has to be approved by the director. Then critics look at the play and write "the direction is wonderful." Garson Kanin would say, "Who can figure out where the play starts and the direction ends."

Do you think you would handle actors differently if you were a director?
Actually, I directed one season at the Metropolitan Opera. Garson Kanin was returning after ten years, and he asked me to assist him. Over the years, I had written down every word he told his opera singers. Every word. So I directed at the Met, while I was playing in *A Funny Thing Happened on the Way to the Forum*. I'm told if you work that hard you're supposed to die. (Laughs.) But I didn't. I loved every minute of it and felt pretty good.

I was also lucky to have worked with some of the best directors that ever lived. Dr. Tyrone Guthrie, George Abbott, Garson Kanin, Gene Saks. I picked up a lot from these people. When George Abbott directed, he moved you.

I did a routine for George Abbott on his 100th anniversary celebration at the Palace Theatre in New York. People write about the glory, romance, and excitement of the theater, yet no one wrote about the paralyzing fear of auditioning for the great George Abbott. I parodied that in my routine—recreating the stark fear I felt auditioning for *Once Upon a Mattress* and *A Funny Thing Happened on the Way to the Forum*. Abbott thought the routine was funny. The fact is, he created fear, but if he hired you, he accepted you for the actor you were.

What about your experience with film directors? Let's talk about the actor-director relationship in Cocoon.
Ron Howard was a good director for me on *Cocoon*. He was a young man, but he knew how to handle some veteran actors like Hume Cronin, Jessica Tandy, Maureen Stapleton, Don Ameche. Now if he made a wrong move, these people would have chewed him up. They happened to be nice people who knew what they were doing. And he let them work. I think it was his perception and gentleness that added to that beautiful flavor of that movie.

You improvised the last scene in the film Cocoon. *Can you tell us about that?*
Yes, the revised scene was much more credible for my character. The original ending called for me to join my friends in a spaceship, after my

wife died. We were all going to go to some interplanetary place. My character, Bernie, was supposed to walk to the spaceship with a bag of clothes and tell his friends to save him a seat by the window.

Before we got to that scene, I new Bernie very well. I told Ron Howard the character wouldn't go on that spaceship. He asked why, and I explained. "Bernie's wife died, his best friends are gone, he's got nothing. He wouldn't go." He's a good director, thought about it, and said, "OK, we'll shoot *two* endings."

We shot the original ending, and then we did the revised ending. This time Bernie apologized to his friends, "I belong here, this is my home." Then I walked pensively away from the spaceship. I didn't know they used the improvised ending until I saw it in the movie theatre.

But they cut some of my best acting. I had a long, lonely walk down the dock until I heard "Cut!" My back was to the camera, and I was thinking intently, "Should I go with them or not?" After seeing the film, I realized I should have turned *toward* the camera so the audience could see me deliberate and think. It was missing from the scene.

Does the writer have any problems with improvising and changing the script in production?
It's unfortunate, but writers have very little to say on the set. One writer I remember on the set of a film was Dalton Trumbo. We were shooting *The Fixer* in Hungary, which was based on a true incident. The dialogue was quite long and expository, it took too long to unearth the story.

Dalton Trumbo is a great scriptwriter, but the director didn't do justice to the writer's intentions. He emphasized the gory aspects of the film rather than human values. Dalton Trumbo was on the set and he made some changes, but he simply didn't get the right break.

Do you find television directors easy to work with, or are the time constraints problematic?
Again, it depends on the person, although there's very little time to rehearse in television in film. I did a pilot, "Trapper John, M.D.," with Jackie Cooper directing. We shot it in a hospital outside of Los Angeles. I played a man who ran away from his daughter and son-in-law, because they wouldn't put him in a nursing home, and he *wanted* to go.

My character was in the hospital bed, had a heart attack, and my voice was very faint. I was lying in bed and Jackie Cooper's face was right above me as I did the scene. After I finished, I asked, "Was that too soft?" He smiled reassuringly, "Keep it as soft as you'd like."

I appreciated that. As an actor, if you think your choice is right, but the director differs, it can change the whole texture of the scene. I liked him as a director.

When a director wants to do several "takes" of a scene, do you prefer to do many or few?
Some actors like to come in, get them over with fast. Others will want the camera to roll and stay rolling. I wouldn't mind doing forty takes. Some people say they do their best work in the early takes. I scrape the barrel, try to do as many as I can. After the director is satisfied, I still ask if I can do another. The director's reasoning may be connected to film costs, not artistic concerns.

For me, the more takes I do, the better I become. With more takes, I *know* more about what I'm saying. There is so little rehearsal time in film, each take gives me a chance to find the right moment in a scene. Mike Nichols only gave us two weeks to rehearse *Catch 22;* Ron Howard had very little rehearsal time on *Cocoon.*

You've played a number of memorable character types over the years. Has typecasting been a problem for you?
I've been very successful in Jewish parts, but I don't want to be type-cast. I've turned down a lot of Jewish parts because once you've played a good Jewish part, everybody sends you the same script. It's always the same character in a slightly different locale and setting.

When Fred Silverman was head of programming for CBS, he offered me two completely different parts for potential series. I was doing Cracker Jack commercials, which were very successful, and he wanted me to do a series based on the Cracker Jack commerical. (Laughs.) I pointed out that the Cracker Jack character was different each commercial. He said, "Exactly." So they wrote a script in which I did something different each week. But the other network executives didn't buy it. They thought a series needs a character the audience can identify with every week.

With that project abandoned, Freddy sent me a new script to consider. It was a pilot script about a retired Rabbi whose wife had died, a young Rabbi comes to town, we clash a bit. I didn't take it. Do you know why? Because if I played the Rabbi, and the series was successful, I would never be offered any other part than a Jewish part.

Freddy Silverman couldn't understand that. He kept calling me, setting up meetings, trying to get me to play that part. When I was doing *No, No, Nanette* on Broadway, he came backstage with his fiancée. We were going to discuss the series prospect over dinner. When we went out the stage door, a lot of people noticed me and came over for autographs. When we got to Sardi's Restaurant, it was the same thing. People recognized me, wanted autographs. All this fed into Freddy's determination to get me to play the Rabbi. He begged me again to read the script. I did, and I still felt the same way.

Months later, I received a call from Freddy Silverman asking to meet him at his office. Some "big guys" from the West Coast wanted to meet me. "Is it about the Rabbi?" "No." So I agreed. His office was on the thirty-second floor of CBS. When I walked in, Freddy was at his desk like a quarterback on the football field, surrounded by four or five other executives. He tells me that the Rabbi is a great part, the series will be very successful.

I didn't have my agent with me, didn't have my lawyer. The only thing I knew was I did *not* want to be stereotyped because of the series. They told me the series could make me a millionaire. (Laughs.) I wouldn't mind being a millionaire, but I leave all that to my business manager.

On the way out, one of the other executives, Perry Lafferty, took me to the elevator and commiserated, "Too bad you didn't take it Jack, because it'll be a whole new ball game next year." (Laughs.) He knew the game well. It's called musical chairs, and it throws out TV executives and their new series ideas every season.

What do you look for in selecting the next project to work on?
I'm looking forward to doing something that can completely use my comedy talents. The closest possibility is the part of Jethro Crouch, a wonderfully stingy 87-year-old character in Larry Gelbart's *Sly Fox*. I did it on Broadway and in Los Angeles with George C. Scott, and I'd love to do the film version.

I'm lucky that I can pick the characters I want to play, and turn down parts that are not as challenging. If I think a character is unsavory, I simply won't consider it. I'd consider the sequel to *Cocoon*, because I love the character and enjoyed working with the cast and director in the original. Normally, though, I wouldn't have great confidence in sequels. Somehow, the elements that made it successful the first time are not there anymore.

I do remember a situation in which I committed to one television film but was overwhelmed by another part that came along. NBC-TV wanted me to do a film called "Hostage Flight." I accepted the job. The money was fine, the part certainly was not complicated. I was a holocaust survivor. In the story, the passengers captured some terrorists

Jack Gilford in his favorite comedy role, Jethro Crouch, in the Broadway production of *Sly Fox*. (Courtesy of Kramer-Jos Abeles Studio)

who were brutalizing them. The hostages wanted to execute the terrorists. My character objected on moral grounds; if we killed them, we'd be lowering ourselves to their level. The character had courage and spirit, but wasn't very complex.

At the same time, producers for PBS (Public Broadcasting Service) offered me a script called, "Seize the Day" by Saul Bellow. As soon as I read it, I wanted to do it. It was a wonderful part. Unlike most scripts, this was filled with warm, sweet people, and I truly loved the project. My character was a colorful phony, a cheat, he knew everything. That's the part I wanted.

I asked my agent to please get me out of the NBC film, so I could do this one. She agreed, and the producer for "Seize the Day" wanted me to start rehearsals with Robin Williams right away. I called my agent on Labor Day weekend to be sure she took care of the NBC commitment. But she was away, no one knew where she was.

So I had two jobs at the same time! I was really frantic. Rehearsals for "Seize the Day" started in New York, and I began rehearsing, though I still wasn't out of the NBC film. Finally, I got word that NBC hired Jerry Stiller to replace me, and he was very good in the "Hostage Flight" production.

It's fortunate I didn't lose *both* parts during that tumultuous time. But that's the choice you have to make if the right part comes along.

There must be times when you are "between jobs" by choice or circumstance. How do you handle that?

In between projects something will come up. It always does. To make sure I stay active, I've worked up a one-man show. It's all about Jack Gilford, with audience questions and answers. I started it in Pasadena, and it's been very successful. It's also a great opportunity to keep tuned to the immediacy of a live audience.

What advice would you give new actors who are trying to break in?
It's very hard to break in. There's no more vaudeville, so you've got to find your own way. You can study acting at school, and you should study about *people* so you understand the way we behave. My wife, Madeline, teaches a class for young actors. In it she explores how to look for a job in film and stage. She talks about the value of imitating to get started, to get a clue of what to do.

Observing and imitating helps a lot in acting. Looking at people—understanding people—is a great resource for actors. I look at people, *all* actors look. Paul Newman said he regrets he's so well known that he can't unobtrusively join a crowd of people and watch them, listen to them. We're all collectors of behavior.

If your interests lie in comedy, try improvisational groups. There are a lot of improvisational groups around, and it's a great way to sharpen your skills of mimicry and imagination. Local comedy club owners might be interested in putting you on for free. Convince them you have the right stuff, and you'll gain the experience of working an audience.

If you've got enough nerve, chutzpa, and talent, you'll find an audience. That's exactly what happened to Robin Williams. It's not inconceivable it could happen to you.

Chapter **9** .

Robert Guillaume

Robert Guillaume discusses how he moved from opera to acting, and how he developed characters for stage and television. He details the process for creating "Benson," his best known television comedy character. He stresses that comedy must be based on truth—in the character, script, and real world. He talks about the need for "improvisational flexibility" in acting. After starring in network series for seven years, he reveals what it's like to go back into the freelance acting world. Robert Guillaume was interviewed at a Hollywood restaurant.

．　．　．　．　．　．　．

Career Profile   Robert Guillaume has performed on Broadway, television, and film. He is the recipient of two Emmy Awards for television comedy acting. His work in musical theatre earned him a Tony Award nomination. He's also appeared in entertainment showrooms throughout the country.

Mr. Guillaume attended St. Louis University, where he majored in business. His ambition was to be the first black tenor at the Metropolitan Opera. After studying voice with Leslie Chabay, he received a music scholarship to Aspen, then apprenticed at Karamu Theatre, where he performed in operas and musicals.

On Broadway, Mr. Guillaume played the lead in many shows, including *Purlie, Golden Boy,* and the Tony-nominated role in *Guys and Dolls.* Among his other New York stage appearances, *Tambourines to Glory, Othello, Porgy and Bess, Apple Pie,* and *Jacques Brel.*

In the early 1970s, Mr. Guillaume was signed to play the role of Benson, a butler in the comedy series, "Soap." His character was so successful, it generated a spinoff series, "Benson," in which the character was elevated to a lieutenant governor. He starred as Benson for the duration of the series, winning two Emmies for Best Supporting Actor (1979) and Best Actor (1985).

Throughout the 1980s, Mr. Guillaume was seen in motion pictures and television films. He starred in Neil Simon's *Seems Like Old Times,* NBC-TV's "The Kid With the Broken Halo," and "The Kid With the 200 I.Q." In the latter project, he also served as executive producer.

Mr. Guillaume continues working as both actor and producer. In 1987 and 1988, he served in both capacities for ABC-TV's "John Grin's Christmas," was executive producer for NBC-TV's "The Fantastic World of D.C. Collins," and starred in New World Pictures' *Wanted Dead or Alive.*

•   •   •   •   •   •   •

Your background and training was in music and opera. Did you have any formal training in acting? Who influenced your performance style?
In college, I studied business and took a lot of music courses. I broke into the business as a singer. If I hadn't been able to sing, I wouldn't have been able to get arrested as an actor. I had a lot of ego, a lot of arrogance, but I didn't have much acting ability. Opera was the only training possible for a singer, and that's separate and distinct from acting training.

I remember wanting to pattern myself after people like Sammy Davis, Jr., Sidney Poitier, Harry Belafonte. As an actor, Sidney Poitier was a major influence. His performances were so real. Later, George C. Scott impressed me. I liked the way he approached what he was doing. All of his roles seemed so natural.

I was impressed by actors who were natural in their performances. They didn't appear to be doing anything, just were themselves. They could go from speaking to acting in one fell swoop. It was intriguing to see that kind of talent at work.

When you get a script for a new part, what's the first thing you do? How do you analyze it and research the character?
First, I'll read the whole script about six or seven times, before I open my mouth. I resist trying on the lines, or any ideas about the character. After I've read the script, I might try out a few lines. Then, I'll

look at what other people say about my character. I'll read the script again to get nuances and feelings that can't be discerned in one reading. I pretty much know the psychological and emotional terrain.

If I need more background information, I'll do research. Sometimes I'll do historical research first. But, I haven't been fortunate enough to get many scripts that require that much research.

A few seasons back, I did a movie (*The Kid from Left Field*) in which I played a baseball player who was down and out. There wasn't a lot of preparation needed. If I were playing an historical figure, obviously, I would have to do a lot of research. But, just to play a baseball player who's down on his luck doesn't require research. It requires understanding of that person in that situation.

Even though I don't do much research, my characters are totally believable. That's what matters to me.

When do you begin to build your character physically?
I use rehearsals as a growing process for my character. The first few times we go through it, I investigate. I find out whether what I thought is actually the case. Sitting down, standing up, moving about.

It's like a kinetic experience. I don't try to push, I let it happen. The physical growth of the character has an evolutionary life of its own. I resent being pushed, and I've found that some autocratic directors don't understand that process. I can't force it. I can't push it. If it's going to be there, it's there.

I like to meet the writer's intent, and within that find some improvisational values for myself. I want to get some "jazz" into what I'm doing. I look for some character traits and attitudes that might be a little off the beaten path, but are still valid.

Give us an example of how you created the character of Benson. What traits and attitudes were you working from?
From the beginning, I understood the character that was created by Susan Harris for "Soap." I also understood the character in a broader sense, a more poignant sense. As an actor, I brought both dimensions to the part. Specifically, I brought a sense of comic outrage, in addition to the brilliant foundation Susan Harris had already given us.

My underlying attitude was a sense of comic outrage black people felt in this society. That sounds heavy, but it was, in fact, built into every line and reaction. That attitude was ingrained in how I portrayed Benson. It was recognizable, understandable for the audience. I think they also found it truthful. That's what made it funny.

Before that role, people would have said, "You can't do a role with that kind of anger. You can't be a black person with that point of view. Nobody's interested in how you feel about society." But I found the opposite. The audience understood that it was a reasonable portrayal. I played a person who was his own man. He didn't care where the chips may fall. I thought of that as truth.

You created a starkly different character in Porgy and Bess. *Tell us how you approached the character of Sportin' Life in that production.*
It was totally different from anyone else's approach. I always felt there was an inaccuracy about black characters that somehow got into that work. I disagreed with portrayals that I saw, because they weren't true. They had nothing to do with real life. Most of the portrayals were demeaning. Or just plain buffoonery. So I deliberately went against that style.

It's difficult to talk about it, but I do remember playing that role completely against the grain in my attitude. I thought of Sportin' Life as an upstanding cat who knew exactly where he was going. If he could take Bess with him, then fine. This cripple Porgy, who the hell was he! Sportin' Life tries to make her see that reality, "Bess, I can't see what you hanging around this place for. With your looks, and your ways with the boys, there's big money for you and me in New York." I mean, that's a rap that's heavy. It's not a dumb pimp's rap. It's a smart rap.

And, even when he asks her to take dope, there's something behind it. "Come on, how about a little touch of happy dust for old times sake? Come on, gimme your hand." All of it made sense. And, instead of being something that simply titillated the audience, it was something that made them think.

The traditional way to do Sportin' Life had been vocally, singing very high: "Picnics is alright, but it's small town, suckers. But we is used to the high life, you know. You and me, we understand each other." There was a kind of other-worldly, snake-like quality about the character. I found it reasonable, so I just sang it low, in my regular voice. That's when I give Bess the rap about the big money in New York.

I found a whole other life for this character. Sportin' Life had a dream of going to New York and making it big. That's sensible. Rather than being a simple, evil person, I found that he was kind of smart. He'd think, "If the rest of these idiots want to stay down here in catfish row, if they think that's what life is about, they're mistaken."

That made a hell of a lot of sense. So, rather than being a clown, he was the smartest man around there. In some ways, of course, he was a snake and scurrilous, but still there was an underlying intelligence. I found myself constantly playing against the grain in that character, which made it very interesting for me and the audience.

You did the musical Purlie *on Broadway and on the road. Can you take us through the process of creating that character?*
Well, I didn't create the role. Cleavon Little had the part before I did. He had many wonderful things, but I didn't want to do the role as a carbon copy. I didn't want to imitate his performance. (Laughs.) I had no shame in stealing some of his better things, if they fit me. But for the most part, I immersed myself in the role. I tried it on. If the left sleeve was too long, I pulled it up to conform with the right sleeve.

I did the last five weeks on Broadway of the original company, then we went on the road a year. When I came into this role, it was well established. It was also a musical, so it had a lot of beats established by someone else. In that sense, I had to replicate what was done to accommodate timing for the other actors.

I don't remember really doing anything original in that piece until several months into the run. The first three months I had very little nuance in what I was doing, everything was yelling, I was constantly hoarse. It took quite a while for me to relax with Purlie. Once I did, it was worth it. I had a good time with that role.

Do you subscribe to any particular school of thought about acting?
There should be no difference between "acting" and "being." I try to eliminate the line of demarcation. I simply approach what I'm doing "as if" it's real. What would happen "if" such and such actually happened.

I've worked with Method actors, and personally, I find the technique boring. For me, the Method fails if it can't produce something the audience feels or experiences. I'd get bored, even in the hands of someone who is supposedly a great Method actor.

After playing television comedy for so many years, what are your thoughts about comedy style?
My concept of comedy has changed over the years. I used to think comedy was more physical than verbal. It had something to do with a dash of buffoonery, a dash of funny movement, not much thought behind it. After all, intellectual comedy was English, so called Restoration comedy, drawing room comedy.

But, after playing comedy for so long, I realized that comedy has a grander purpose. It brings out a heightened sense of truth. It's going for truth, rather than physicality.

My best known character was Benson, and the only thing I searched for was a sense of accuracy. I didn't have to be funny. I just had to say what I'd been charged with saying. If it was funny, fine. If it wasn't funny, that was fine, too. I never felt the need to be funny. What I said was funny, how I acted or reacted was funny. And it was funny because there was always an element of truth behind it.

What is your ideal relationship with a director?
As an actor, I want the latitude to try several different tacks in a scene. A good director will understand that. I don't want to stick to the same reading all the time. There are thousands of nuances within any given line. Nuances of attitude.

When I find I'm with a director who wants me to do it as I did it before, that frightens me. Because I don't think in terms of repeating a performance. I want some improvisational flexibility.

Remember, we're not talking about vast swings here. We're talking about little micrometers of changes. And that's all I mean. If a director wouldn't let me do that, we'd have to part company.

How do you handle the conflicts when you and a director disagree?
I try to reduce any tension on the set, especially any sense of competitiveness. We're trying to help each other, and we're helping the other actors. The more comfortable the director is, the greater my chances of being able to function freely as an actor in that environment.

Obviously, I wasn't hired to be a director's puppet. I'm expected to deliver something creative. If we both agree on what that is, things will work out fine. If not, I'll get a whole slew of notes.

With director's notes, we start talking too much about the script. That bothers me, because there isn't a hell of a lot to talk about. Anyone can figure out what the writer meant. The challenge is—How do we show it? How do we demonstrate it? That's what acting is about, demonstrating, showing. It's not about intellectualizing or philosophizing. What we're trying to do is find out "how," not "what."

The craft of television encompasses work with and without audiences, shooting multiple takes of scenes, and shooting out of sequence. Can you discuss these aspects of your work?
Most of the comedy shows we shot were in front of a live audience. We did that for many years on "Benson." Sometimes the audience took

over, telling us what the scene was about. And it may not be what we had in mind. We'd have serious dialogue, and yet someone would laugh. I'd wonder, how could anybody laugh at that point? The audience had a life of its own, it was unpredictable.

As a change, we tried a few shows with no audience. It was a refreshing process. It was actor to actor. No outside intrusions on those moments. It was interesting, challenging.

As for different takes of scenes, I like to do the scene several times. It increases my sense of spontaneity. I don't want to have my words so well memorized that they sound memorized. I like to be a little bit on the edge, not knowing what I'm going to say next. It adds a texture to acting.

You asked about shooting out of sequence. That bothers me. I would prefer to shoot in sequence, because it's difficult to know precisely where I am. Long after I've done a scene, something will hit me, "Dammit, I was supposed to do this, but the scene before I did that." You can have nightmares about whether it's going to fit together, emotionally, dramatically, progressively.

How do you feel about the casting process?
I never enjoyed being rejected. Being told, "No, you're not the person we want." I never enjoyed that. But I always liked the challenge of auditioning. When I first came into the business, I could sing pretty well. I had enough training to know how to make a song effective in an audition. That's a long way from professionalism, but still, it got me in the door and kept me there for a minute or two. When I auditioned I always got the part on sheer talent.

The problem I have now is that with a measure of celebrity, people assume I'm no longer interested in the casting process. And so there are perfectly good parts that I would like to play, that I never get an opportunity to even look at. People assume, "He wouldn't do that."

You starred in a television series for seven years. Now you're back in the freelance talent pool. How do you feel about that kind of instability?
I can deal with it better than when I first started out, because I have a little money, and some of the accouterments of success. In that sense, it's quite unlike the days when I was in New York and didn't know where the rent was coming from.

But there is *still* tremendous frustration and insecurity. Despite all the success and the financial rewards, I still worry about the next job. As soon as one job's complete, I'm feeling, "What the hell is next? What am I gonna do? Will I have to go work at the post office?" Even when I was on the series, I'd try to keep busy during hiatus periods. That's when I'd do benefits or schedule other engagements.

People tend to think they've seen the best of you if they've seen you in a series for several years. They don't realize that what they've seen is simply a structured idea, and you've stayed in one place. There's a lot more potential that can be tapped.

As an actor, you want something to do. And it's always ironic that once you become a star, people think you wouldn't be interested in doing different parts. There's a lot of things you would do, because it appeals to you. It doesn't have to be the lead, it doesn't have to make a big splash.

As an actor, I *want* to do a lot of different things. I'm an actor. I'm a worker. So, let's get to work here. Actors relish the work process.

What advice would you give to a person interested in acting, particularly with regard to training, breaking in, and surviving as a working actor?
It depends on where they are in time, where they are in life. To beginners I'd simply say, take classes, observe life, people, as much as you can. Try to reduce the number of preconceived notions you have about this craft. Understand that it's called "show business"—some "show" and a *lot* of "business."

For people more experienced, I'd say try to reduce the level of anxiety about your career. Put yourself where things are happening. Do everything you can to increase your knowledge of what's going on. Do the leg work that's required to make it happen for you. Sell yourself. Keep that firmly etched in mind, you are selling yourself.

I used to sell pots and pans before I got into the business. I wasn't a good salesman, because I didn't give a damn about pots and pans. I was doing it to earn a buck. My supervisor would come around with promises of unlimited income, and that motivated some salesmen. He'd say, "If you want one sale a day, you've got to go out twenty times." That meant being rejected twenty times.

And it's something like that in our business. I didn't realize that as an actor you have to sell yourself. People often have difficulty with that notion, but it's important. Do whatever you can to get the job. Sell yourself.

When you go on auditions, you can sell yourself by being relaxed and in control. Sometimes the process, itself, is unnerving. The audition scares you, and before you know it, you're working on some kind of energy that's counterproductive. That energy pushes you past where you want to be in some areas, not enough in others. You have to be able to concentrate on exactly what you think, and turn the screws.

Try to relax. Try to increase your concentration, so the auditioning process doesn't rob you of what you did to prepare. The approach you've worked out in class will certainly be effective. You can enhance your ability to concentrate on that when auditioning.

Another thing to keep in mind. Acting is not likely to make you a lot of money. In that way, actors are like other artists and craftspeople. Poets don't make money, but that doesn't stop them from writing poetry. The same is true for all people who want to create.

What do you think it is that allowed you to beat the odds in the business?
It had to do with "luck." There are a lot of people who are talented, but that's not the key. Talent is certainly a part of success, but luck is vital.

I was just as talented in New York, though I may not have known it at the time. Still, I put myself out there to get roles. A lot of parts I played were already established. The blush was off the rose by the time I did it. I did Jacques Brel for about eight hundred performances, and people still remember it. They'd say, "I remember seeing you in Chicago. Boston. Washington, D.C. Los Angeles. I never had a night in the theatre like that." But I didn't originate that role. In a sense I was able to hone my craft from it.

The luck I needed, I didn't encounter until I got the series, "Soap." Then I was noticed for the first time. For some reason, the public took notice of the character, and loved seeing him do crazy things on national television. It hit.

To me, that was serendipity, and luck. That's all it takes.

C h a p t e r **10**·

Bart Braverman

Bart Braverman talks about character preparation and timing in comedy, building characters through script analysis, and dealing with the pragmatics of television production. An actor since he was five, he candidly discusses job cycles in acting and imparts advice about dealing with employment instability in the craft. Bart Braverman was interviewed at his home.

. . . .

Career Profile Bartley Braverman, a child actor in the 1950s, appeared in numerous television productions. He was seen in all the classic television programs of the period, including "I Love Lucy," "Have Gun Will Travel," "The Jack Benny Show," "The Red Skelton Show," "The Bob Hope Show," and many others.

He left Los Angeles to study at Carnegie Mellon University, where he majored in drama. He studied Shakespeare and contemporary musical theatre, then landed a role in the rock musical *Godspell.* After eighteen months on the road, he returned to Los Angeles, starring in a cult musical, *The Rocky Horror Show.*

Although he had a busy career as a child actor, acting jobs suddenly became scarce. During that period, he worked in production with his brother, filmmaker Chuck Braverman, and joined an improvisational comedy group in Los Angeles, "The War Babies."

Mr. Braverman performed with the Los Angeles Shakespeare Festival for two seasons in the mid-1970s and was cast in several television pilots. He starred as Binzer in the ABC-TV's series, "Vega$" for three years (1978–1981). He later starred in the comedy series, "The New Odd Couple."

Throughout the late 1980s, Mr. Braverman has guest starred in many episodic television shows. Among them, "Murder, She Wrote," "Simon and Simon," "Matt Houston," and "Shadow Chasers." He also appeared in television films such as "Prince of Bel Air," "The Gladiator," and a pay cable cult favorite, "Alligator."

. . . .

You've been acting since you were five years old. Obviously, you couldn't have much training at that age. How did you learn the craft?

When you ask how I learned the craft, it's almost like asking, "How did you learn to breathe?" I've always acted. To me, it's one of the most natural things in the world. I can't really remember a time when I wasn't acting. My mother was an actress, and it was simply a part of our life.

When I was five, my mother's friend was doing the casting for the Ed Wynn show. He needed some children for a Pet Milk commercial. My brother and I were picked out of hundreds of kids. The commercial was set in an old western saloon, and all the cowboys were small children. That was my first professional job.

As a child, I didn't have any formal acting training, just worked from experience. When I was twelve, I acted in a stage play, *A Hole in the Head,* with Jesse White. Other than that, most everything I've done was in television and film. I learned a lot from doing it.

It's ironic; when I went to college, I wanted to study for a career outside acting. I wanted to have another means of making a living, something to fall back on if my acting career didn't pan out. Instead, I majored in drama right away. It's where I felt most comfortable. We studied a lot of classical theatre and Shakespeare. The first year I played Iago, next year I played Hamlet. That's where I "formally" learned the craft of acting, in college.

When you were ten years old, you were cast as an Italian shoeshine boy in "I Love Lucy." How did you prepare for that, with accent and all?

I've seen that episode many times in reruns. The shoeshine boy reminded Lucy of her own son. Actually, I just played the lines for comedy. I knew what was funny, and *why* it was funny. I understood comedy. As for the accent, I could do a pretty good Italian accent, and that's what got me the role.

Many people still ask if that episode was filmed in Italy. It was one of a dozen shows in which Lucy, Desi, Fred, and Ethel went to Europe. The stories centered around Ricky's band touring overseas. Everyone thought we went to Europe to film, but the fact is, we never left Paramount Studios. We shot it right there on the stage.

You've worked in live television with comic giants: Lucille Ball, Jack Benny, Bob Hope, Red Skelton, Ed Wynn. What did you learn about comedy?

Two things are essential—character preparation and timing. Character preparation is important, unless you're just doing physical comedy. If you're doing a character with any dimension, find something the audience will believe in. It has to be identifiable. Even in comedy, there's a character inside. You've got to think, what motivates him? What was his past like? What brought him to this moment?

The other element is timing. Timing is critical, because no matter the motivation, the character has to get a laugh. That's the most important objective in comedy, getting the laugh.

In a television series, physical comedy is simpler. You don't need to build much character. You read the script and try to understand what the writer was thinking when the dialogue was written. Why is it funny? Why was this line written? How does it fit in? Your objective is to make the line work; to get the laugh. As long as you get the laugh, you're successful.

What acting techniques do you find most practical? Have you worked with actors trained in the Method?

One teacher stressed something I still use—reading the script carefully. Everything an actor needs to know about a character is in that script. He taught us to look at everything in the text: what the character says; what the character does; what's said about him. All the important information is in the script.

As for Method actors, I often get annoyed with them. My impression is that they take themselves too seriously. I happen to think that good acting is like good painting. It can bring you to tears in a minute. It's an art. But it doesn't have to be so damn serious.

I know a Method actor who is a fantastic comic but is terrible as a dramatic actor. I think it's because he takes himself too seriously. Some

Method actors are just so *full* of themselves. I personally think that gaining sixty pounds to prepare for a part is too much.

That's not to say that some Method actors are not brilliant. Think of Al Pacino, Bob DeNiro. They're great actors. There is no doubt about that.

The reality is someone can walk in off the street and win a part by virtue of "looking right." But that person won't be able to play other character types believably. To play a low life mafioso hit man one day, and a sophisticated constitutional lawyer the next, you *have* to be trained in a more formal way.

I have to admit that while I'm annoyed by actors who take themselves too seriously, I almost envy their passion. They do seem to have a great unbridled passion for their work. I'm afraid I don't have that passion, or at least it's rare.

The closest I've come to that feeling was in the television series, "Vega$." I created a character, Binzer, who had a lot of depth, and I really understood him. But the process was so easy. I got up in the morning, memorized my lines, walked out on the set and did it. It just wasn't hard work. There was something about the character that came very easily to me. There was a lot of him in me, a good many character qualities we shared.

Now, if somebody else played Binzer, I'm sure they would have created a different character. There's no question about it. But the character I created was a lot like me. I'm not too happy about that, since I didn't particularly like those qualities.

When you get a script, what process do you go through to create that character?
The first thing I do is carefully read the script. I want to see what happens to the character. What does he do? What's done to him? What do people say about him? From the script, I learn what's going on in his life, and I try to learn what went on before the scene started.

For example, if the character has a limp, he's battled a physical disability for years. That's an interesting problem, especially if you want to build a complex, identifiable character. You can't just develop a limp on stage. It means you have to ask, "What must it have been like for a boy to grow up with a limp? What must have gone through his head?" And that requires a certain amount of creativity, empathy, imagination.

I don't think you have to dwell on it, or work on it excessively. The information can be processed and used whenever needed. It goes into the mental computer. Once I've got all that absorbed, I go in and "play the moment." I can only play the moment that exists at the time. A person who grows up with multiple sclerosis is not thinking every moment of his life, "Oh, no, I've got multiple sclerosis." Sometimes the person thinks, "I'd really like an ice cream." It has nothing to do with having a disability.

On the other hand, if the same character has to go up to a woman and say, "Excuse me I left my money at home. Can you loan me some money for an ice cream?" He might be dealing with all the women that rejected him in the past. That's in there, it's got to be.

As for actual background research on a character, life experiences are *very* valuable. There's so much you can use. However, that doesn't necessarily mean I can draw everything from my own experiences. If I'm playing a rapist, I can't go out and rape somebody. I can't imagine what would be in that kind of mind. It's so foreign from me that I probably would have to do actual research. I might arrange to talk with a convicted rapist, or talk to someone at a rape crisis center. Maybe talk to a psychologist or psychiatrist about the personality type. It's a hard role to imagine creating.

You've worked in theatre, television, films. What comparisons can you draw about the acting environments?
I think most actors prefer the stage. In theatre, you can turn to a director and say, "That didn't feel right, let's try something else." There is less pressure, less tension, less risk. There's a great commitment to creativity on stage. There's always time to rehearse and get it right.

When you're in front of a live audience, the old idiom always applies, "The show must go on." Athletes endure pain and suffering that they wouldn't go through in *any* other area of their lives. Standing in the coliseum with eighty thousand people screaming, something extraordinary's happening. That same energy applies to acting. When you're on stage, with people out there, and you create a moment, that's *it*. There's nothing like it in the world.

In television, there's too little time to do anything like that. No time to rehearse. The name of the game is "get it done quickly." They're working on a budget, got to produce fast. I'm proud that I have a reputation for getting things done on the first take. I make sure I'm prepared, and I'll help any other actor who needs assistance in other takes. There's no time for rehearsals in television.

With so little time for rehearsal in television, how do you prepare for your work on the set?

It's all done before coming to the set. There's plenty of time to prepare alone, but it's difficult, because you don't know what anyone else is going to do. The fact is, "rehearsal," as we think of rehearsal in stage work, is *nonexistent.*

I asked a director one time, while I was doing "Vega$," "What do you think about this particular moment? Do you think Binzer would play it this way, or that? What do you think?" He shrugged it off, "I don't care. Do whatever you want." Time was pressing, this wasn't relevant to him.

So you have to come in prepared. You have to know what you're going to do. You must also be prepared to change the instant the director doesn't like what you've brought in. It's not very often, of course. There's an old joke about television directors. They're said to be doing their jobs if actors say their lines and don't trip over stage props. Sadly enough, it's more true than not.

I can still bring something to the part, even without attentive directors. Just because they're not taking time doesn't mean I can't be good. It's likely the other actors are fully prepared. They've thought about their roles, their motivations, their actions. As a result of that creative mix, there's a chance to create a lot of nice moments on screen.

Given the problems in television and film production, how do you view the actor-director relationship?
Remember, I was brought up with great respect for the director. That voice is the law, the final word. You can voice your opinion, but once the director says, "I want to do it *this* way," you *must* do it that way. A good director will tell you exactly what he or she wants.

If you don't get along with a director, it will certainly affect your performance. It's a terrible situation, because you are performing to please the director. On the set, that's the only person you're performing for. If you don't get along, your scene will be shot over and over, and you'll be furious, thinking it was perfect the first time.

If you continue to fight, two things can happen. You'll get fired, or you get a reputation for being a pain in the ass. That's all you need in this business. It's hard enough to get a job as it is.

I worked with my brother (Chuck Braverman) in a movie called *Hit and Run.* He's a good director, and I enjoyed working with him. Of course, I also walked on eggshells, since he's my older brother. If I disagreed with him, I wouldn't make it public. I'd take him aside and give him my opinion privately. He respected my ability and my opinions as an actor, but I would never put him in a tenuous position with his crew.

How do you prepare for shooting out of sequence?
It's certainly more inconvenient than working in sequence. You have
to plan how you're going to play every scene in the show. If you're
doing the eighth scene *before* you do the fifth scene, you have to know
what your reaction is going to be in the fifth scene, because it will affect
what your reaction is going to be in the eighth scene.

You have to plan what all of your reactions are going to be. What
all your feelings are going to be. If, for instance, in the fifth scene you
were crying, it would be inappropriate to be laughing in the sixth
scene. But, if you shot the sixth scene first, you don't know whether
you're going to be laughing or crying, how can you play that scene?
You *have* to decide in advance. It can be very confusing if you're not
prepared.

*You were in an improvisational group for quite some time. Do you
like to bring improvisational qualities to your work?*
I love to, but some people aren't comfortable with it. They prefer stick-
ing to every word in the script. If you're in the middle of a script and
say, "How come?" instead of, "Why?" some actors and directors don't
know what to do.

Personally, I feel that being able to improvise on the spur of the
moment, to some degree, is part of the business. If you're doing a scene
and somebody sneezes, it's only natural to say "God bless you." If you
don't, you're not actually in the moment. You're not reacting to reality. What is happening is somebody sneezed; the automatic response is
"God bless you."

Of course, there may be a very good reason why your character
wouldn't say those lines. It might be completely inappropriate. If somebody's holding a gun to your head, and sneezes, you wouldn't say,
"God bless you."

Given your choice, what kinds of roles would you like to play?
If I could choose the parts that I wanted, I would look for the roles with
the strongest social impact. I'm hokey enough to believe that I might
have some impact on people. In fact, the *writer* is the one who has the
real influence. That's the person who came up with the script and the
words. But, to be able to affect someone's life, because of an *acting* job,
there's nothing better.

I'd also like to play fantasy roles. Something like the lead character
in *Ferris Beuller's Day Off.* That character was playing every kid's
fantasy. I'd like my ideal character to arouse a tremendous amount of
empathy and sympathy.

You've been in the business a long time, spent three years doing a top-rated series. The cycles of high and low must take its toll. How do you deal with that?

Not very well. Yesterday, I had to audition for a three-line part; it drives me crazy. The reality is that my career is at a temporary lull, and I have to deal with that. I'm not happy about it. I am embarrassed by it, but it's the reality. And unless I'm going to sell pens, or find another way of making a living, I have to go out and audition like everybody else. The auditioning process is the worst possible way to pick an actor for a role, but it is the process everyone must go through.

All actors *do* go through cycles of good and bad. It's the nature of the business. Bette Davis, years ago, put an ad in the trade papers, "Bette Davis available for work. Someone please hire me." She didn't get hired because it just didn't occur to people that such a star would be available.

Actors do have lows. Big stars have lows. Big stars sometimes go out of fashion. It's difficult. You have to deal with it as best you can.

What advice can you give to new actors about handling that kind of insecurity?

First, don't do what I did. Find another major at college, something other than drama. The best thing you can do, and I mean this seriously, is find another way of making a living. Just because you want to be an artist doesn't mean you have to give up everything for your art. There's nothing wrong with being an accountant, or knowing how to make pizza, or being a lawyer.

I ran into Barry Gordon[1] the other day, who's been a child actor as long as I have. He's been going to law school at night. I have nothing but admiration for that. The point is, if you have another way of earning a living, it greatly reduces the tension.

If you're going to be an actor, prepare yourself for the reality that there are over fifty thousand members of the Screen Actor's Guild in Los Angeles. Only five percent of those are working actors, making a living every year from acting. That's twenty five hundred people. Of those, just a few hundred people are making a *good* living.

Let's look at this another way. There are nearly forty-eight thousand professional actors in Los Angeles who are not working regularly. And among them, there are a lot of very fine, talented actors. Talent

• • • • • • •

[1] Barry Gordon was elected President of the Screen Actor's Guild in 1988.

doesn't equate to success in this business. There is no guarantee of success.

Given the cycles of high and low points in acting, what do you think enabled you to endure all these years?
I really don't know. And that's what's so frustrating. If I could quantify it, I'd know exactly why I'm up or down. If I could point to something specific, it would be helpful. "I'm only five foot nine. If I were six feet, I'd be successful." If I had that kind of objectivity, I could evaluate my career, and say I'll never be six feet, so I'll quit now.

But there is no measurement like that. Mickey Rooney's very short, and he is a big success. Ironically, even his career had highs and lows. He won an Emmy, and in his acceptance speech, he talked about those cycles. When he was nineteen years old he was the biggest box office draw in the *world*. And when he was forty, he couldn't get a job. Same actor, same talent. Couldn't get a job. I don't know why.

I wish I knew the answer for determining success and holding onto it. I don't. It's just part of the insecurity of choosing acting for a profession.

Chapter **11** ·

Jay Johnson

Jay Johnson talks about Edgar Bergen's techniques of ventriloquism, and demonstrates his own technique for creating a new character. He discusses finding clues to characters from vocal and physical behavior and recounts personal experiences in the audition process. He stresses the importance of creating the illusion of reality for audiences and evaluates the artistic and psychological dimensions of ventriloquism. Jay Johnson was interviewed in his studio at home, amidst a forest of puppets.

· · · · · · ·

Career Profile   Jay Johnson is a ventriloquist and nightclub performer who starred for four years in the television comedy series "Soap," playing the dual role of Chuck and Bob Campbell. He's performed in every major venue of the country, including Las Vegas, Atlantic City, Reno, and Lake Tahoe.

As a teenager in Richardson, Texas, he did television commercials with his puppet partner "Squeaky," who later became his current partner, "Bob." He graduated from North Texas State University with a major in marketing and performed in Houston Theme Park shows.

Mr. Johnson moved to Los Angeles to build a television career. He performed in The Horn, a showcase club, where he was seen by ABC casting executives and signed for the "Soap" television series. He also appeared on television talk shows and variety shows such as "Merv Griffin," "Dinah," and "The Jim Nabors Show." He is a regular guest on the "Tonight Show."

Throughout the 1980s, he's appeared regularly on network television, cable, and syndicated shows. He starred in a dramatic NBC Television film, "Riddle for Puppets," playing the part of a schizophrenic ventriloquist. He's guest starred in series such as "Simon and Simon," "The Love Boat," "Gimme a Break," "Facts of Life," "True Confessions," and many pilots.

In addition to performing, Mr. Johnson is now producing shows for television. He's developed and produced entertainment programming for NBC-TV, Showtime, and Home Box Office.

* * * * * * *

Did you have any formal training as an actor? Any early influences on your career as a ventriloquist?

I had acting classes in high school and college. A friend, Wilson Peach, did local television commercials with me in Dallas and made me realize there is a craft to acting. He thought of himself as an actor, and I began thinking maybe I was an actor too. Since then, I studied in Los Angeles, with Charles Conrad, a cold reading class, and with Allen Miller, who is a wonderful teacher.

I was inspired by Jimmy Nelson. But my early heroes date back to the days of Milton Berle Texaco Theatre, and Edgar Bergen, the most famous ventriloquist of all. His characters were wonderful. His technical ventriloquism might not rate among the best in the world. But he was a consummate actor.

Bergen's puppet, Charlie McCarthy, now sits in the Smithsonian Institution. We revere the character he created as an American icon. If Bergen simply created the role of Charlie McCarthy, without the puppet, he would have gone down as one of the greatest actors of our day.

His characters are full of depth and compassion. Charlie McCarthy. Mortimer Snerd. They almost transcended any techniques of ventriloquism. They were beautiful characters. When he devised them, he always started with the inside of the character and worked out. That was his style.

In fact, Bergen had three or four characters that were never seen by the public. He had them carved, but didn't feel comfortable with them. He didn't know who they were. So he never performed with them.

I buy that a hundred percent and use it in my own craft. If someone gives me a new puppet character, and says, "Do something with this," it doesn't quite work. It has to start from the inside out. I want to have it all in my head and then go to somebody who makes puppets and say,

"I want this kind of look, this kind of lips, this kind of mouth, this kind of hair, this kind of eyes. I want him to do this." That's who the puppet eventually is.

Can you give us an example of how you create one of your puppet characters?
I always liked monkeys, so I wanted a monkey puppet. I came up with the character of Darwin, my jazz monkey. The first thing I did was put the puppet on and go in front of a mirror. I just watched him move, as if I weren't moving him. The way he looked at me could be funny. It might be funnier if he cocked his neck and looked up at me. I tried it. And it was a different value if he just looked at me straight on.

Jay Johnson and Darwin, the jazz monkey puppet he created.

Then I realized that a monkey was constantly moving. In a monkey house, that's all you see, a monkey moving. So the challenge was to combine all those movements together. He was never still. That led me to the idea that his attention span is not even as long as a child's. He was going from thing to thing, in constant motion.

From that I could start writing dialogue. His attention span's very minimal, so he's always looking up, getting me to look away. He's looking down at my private parts. I mean, he's a monkey, an animal. In terms of dialogue, he could say almost anything and get away with it. I'm not talking about dirty material, I'm talking about his point of view, which is totally uncivilized.

I also realized he was wild and loud. A monkey screams a lot in the zoo. Once I combined all that, suddenly, he started to live. He was a real character. And where did it all come from? From seeing his movements in the mirror.

Can you discuss the artistic techniques you need for this specialized craft?
I always compare ventriloquism to music and musicians. Anyone can play notes on a violin, but you're not a professional musician if you can just play notes. You're a musician when you know the notes and can almost do them without thinking. Then, within those notes you put the color, the beauty, the actual sound, and the interpretation.

Invariably, amateur ventriloquists come up to me and show me their technique. Their mouths don't move, so they think they've got it down. But they haven't presented a character. They've only given a sound. Moreover, they're usually expressionless. Not a reaction on their faces. That's not ventriloquism.

Ventriloquism is knowing how a puppet works without thinking about it. The voice and character are second nature. I don't think about technique when I'm on stage. I throw that out and let it happen. I get just as amused at something my puppet says as the audience does. I hear it exactly like the audience is hearing it.

On stage, I'm trying to tune in on the audience and tune out on the puppet. I watch the puppet from the perspective of the audience. I key in on them. No matter what, I'm seeing what they are seeing for the first time. If they titter, then I've got to be attuned to that.

The person on stage is not really me. It's close, but it's not me. When I walk on the stage with my puppet, Bob, I become the less witty half. I give into him, to the jokes; I'm really there to set him up. I follow what he does. I react to what my puppet's doing.

I don't consciously think about any of this. It happens instinctively on stage. Similarly, I never think in terms of physically working a puppet. I never think I have anything to do with that, never consciously aware my hand is turning the stick.

The ventriloquist's best mirror is the audience. If they laugh at a specific movement, you'll never forget it. If they don't laugh, you'll never use it again. There's something wondrous about applause or laughter. The minute it happens, it tatoos what you did on your brain, and you never forget it.

You were just given the script for a new project. What's the first thing you do with it? How do you analyze the character?

It's an interesting question, because school never prepared me for that practicality. I learned how to do a character and how to get into a character. But when you get the job, they hand you the script and say, "We shoot next week." You're on your own.

This is the way I do it. I read the script to know the overall story. Then, I read all the scenes that I'm involved in. I know everything that I'm supposed to be doing. I mark my part, and try out some dialogue. Before I have any character, or any idea of where I'm going, I just want to see how the words feel.

If the lines don't feel right, I ask myself why not. I have to analyze it. How can I make it work? What kind of guy speaks like that? How is he different from me? What would I say? Why does he talk that way?

A lot of times, that's the key to a character. Just the fact that the words don't feel right. Then I have to investigate his character. Why does he speak that way? He may have been educated at a certain college, or he has a certain job. I can make up a text if there is none. I can flesh out his background and make him credible.

What about giving the character a physical behavior and a voice?
You start with a walk and a voice. That's how somebody once described approaching a character. And I agree with that. In working with words, I analyze how they sound. By doing a voice, it becomes natural. I might change my voice a little to accommodate the words.

As far as the physical aspect is concerned, I try out the walk. I get to know the character through the way he walks. I try it all on. The externals have to work, and they have to agree with my internal instincts.

If you're supposed to be gray haired, they're going to put you in makeup. But, you have to feel like a gray-haired man. A gray-haired man feels different inside than a guy with no hair, or a guy with some other kind of hair. Unless you know how that feels, you'll just be an actor in a wig.

You played the role of a schizophrenic ventriloquist in the television film, "The Riddle for Puppet." How did you research that part?
That was a particular challenge. I remember doing a lot of research. This guy was a ventriloquist, so I knew what he did. But, this guy also kills a person. I've never killed anybody. And he killed a guy in a very specific way, in a wood workshop.

They took a photograph of the workshop where most of my puppets are made. The man that created puppets for the film, Rene, was in there. I looked around, trying to figure out what weapon I could use.

The script was unclear about it. So, I remember coming up behind Rene, asked him what would be the best weapon within reach.

He didn't hesitate, picked up an awl, which is an ice pick. It's a very wicked looking instrument, and he handed it to me. In the final production, it became a chisel, which is not as good a weapon. I remember thinking, "What would it be like to really want to kill?"

The character had to do several things that I just don't do in my life. In addition to murder, he had to treat the puppet very roughly. There was one scene, very hard for me, in which he smashed a puppet against the wall. The puppet was made to smash, but it was still one of the hardest things for me to do. I'm not trained to beat up a puppet. It'd be like Izak Pearlman breaking a violin over somebody's head.

With a schizophrenic role, there's going to be good and evil. There's a duality of thinking. I had to maintain my good parts and the bad. If either was missing, the characterization wouldn't work.

I approach drama and comedy differently. Comedy is easier. I can see the beats and instinctively know how to make something funny. It's a rhythm. Drama is more difficult for me. I have to work harder to find the dramatic subtexts and play them out.

What is your ideal relationship with the director? Do you like a strong director?

The director is the kingpin on the set. For that reason, I like strong directors. They can take actors anywhere, give them the right situation and make them, at least, do a passable performance. The way they shoot the film, and cut it, will work. But, a great performance is really going to have to come as a team effort, and it's weighted heavily on the side of the actor.

Even if you don't agree, at least you have a definite statement from a strong director. If a director is unsure, you can do anything you want, but you'll never know if it's good, bad, or indifferent.

In a comedy, I once played the part of a bad ventriloquist who moved his lips. The director wanted me to mouth every word, he thought that would be funny. I felt the character wouldn't be so blatant. Even if he was a poor ventriloquist, he'd at least try *not* to move his lips. We shot it two ways, and the director saw how uninteresting it would be for me to play it so broad. Finally, he was convinced I had a point.

He approached actors all wrong. The minute we sat down at the reading, he set down these ground rules, "Don't anybody raise their eyebrows. If you do, I'll yell 'Cut,' and we'll redo it." (Laughs.) Now, I had never thought about raising my eyebrows to do a line. But for the life of me, every time I read a line, he'd yell at me for raising my eyebrows.

With all the changes in television and film, directors tend to be more technically oriented. In the old days, they were actor-oriented. Many of the directors coming up are excellent at catching something on film, or tape, but when it comes to giving you an idea or helping you along, they're not equipped. They don't know how to help you.

How would you compare the performing environments of television, film, and stage?
In both stage and nightclub work, everything is immediate. It's you and the audience. That's the unique satisfaction. An actor can feed off the audience's energy.

My absolute favorite form is film. Single camera film. On the set, there's only you and the crew. When everything gets quiet, it's just you and them. No audience, no waiting for applause. You feel everybody's energy on a set, helping you out. It's a much slower pace. That gives me a lot of time to uncover new information about my character.

In film, you have to deal with multiple takes of scenes, different camera angles, and out-of-sequence shooting. How do you handle those production pragmatics?
In one of my classes, they said the ultimate challenge to an actor—the *ultimate* challenge– to prove that you knew your craft would be to shoot a scene in close-up, with no dialogue. You said nothing, but you had to be alive. You had to listen, you had to react, you had to feel, you had to go through all the emotions. Close-up acting allows that discovery of character.

I don't enjoy doing a lot of takes, but I do like different camera angles and interpretations. The slow pace permits that kind of experimentation. They shoot a master shot, which is the wide angle that establishes the scene. Then they reset and relight the scene for a new angle. It could take an hour to set the lights for your close-up, or two hours if they're going to another person's close-up.

With all that time, I have the chance to do some things totally different in the close-up. I stay within the dialogue, the timing, and the rhythm. But I have a chance to do some things I just discovered.

As for shooting out of sequence, I've been fortunate. The easiest scenes are generally shot first, the more difficult scenes, later. I'm not sure what I would do if I had a death scene on the first day. (Laughs.) Then I wouldn't think out-of-sequence shooting was such a good idea.

How do you handle auditions and cold readings?
I'm dyslexic, so I'm not a great cold reader. I'm concerned about saying the right words at the right time from a piece of paper. If the part I'm reading for is close to me, it's that much less I have to worry about.

I feel confident if I'm up for the part of a ventriloquist. I walk in with my puppet, and if I'm the type they're looking for, I'll get the part. But I have difficulty with auditions if it's just me and a few other guys who are my type.

I recently was on one of those cattle calls. Actually, it wasn't a cattle call, since it was only four guys. But, four similar types, competing for a small role in a feature film. Before I got there, I reassured myself, "This is not going to make or break my career. It would be nice to do a film, but it's no big deal."

When I got there, I could see these guys really wanted the part. They were getting very nervous, and the adrenaline started pumping. If it's that important to them, it should be more important to me. What's the matter with me? This is competition.

And then, the worst happened. All of us could hear the audition in the next room. The part called for screaming and yelling, so we heard the screaming and yelling. As soon as one actor raised his voice, those waiting in the outer office gave each other the knowing eye, "That ain't it." And the other guys agreed, no matter what. (Laughs.) I knew in just a few minutes, they'd hear me yell, and they'd look at each other pitiably and all knowing.

When you were doing "Soap," you had five months off each season. How did you feel about hiatus periods?
With a successful series like "Soap," I knew I'd have a job in the fall. It was only the last year that we were on the edge. We were always sweating a network pickup. The last year or so, I worked particularly hard during hiatus, kept the fall open for nightclubs. Just in case the series wasn't picked up.

I never thought hiatus was "vacation" or "time off." I always looked at it as if I might be moving into the next phase of my career.

What about the inevitable lull between projects? How do you deal with that insecurity?
I don't deal with it well at all. I've never dealt with periods of unemployment, which we all have. I've just gotten off the road from a tour with Julie Andrews. It was a great tour, six weeks long. But I was ready to get home, wanted to give myself two weeks off.

Now I feel the anxiety. (Laughs ironically.) "Will I ever work again?" "Was that my last tour?" "Is my career over?"

I've never found a way to deal with that comfortably. I've never felt good when I'm off. Either I'm not working, or I'm not working enough, or I'll never work again, or I'm always working.

I don't know how to come to grips with that.

What advice would you give to people who would want to pursue a career in this craft?

If you're interested in ventriloquism, understand it's different from traditional acting. As a student in high school and college I did a lot of acting. I worked on scenes with a partner. That experience is vastly different from working with a puppet. It's perceived as similar, but you have to react within the same second and sound. It's a unique challenge.

Advice to ventriloquists. Advice to actors. Advice to anybody. Don't feel like you're in competition with others. Find your own mark. Try to make what you do better. Have no other goal, not a million dollars, not all the things that go along with being successful. Just be good at what you do.

When you go on auditions, try not to think of it as competition. Don't let the competition get to you. If you're really destined to do that part, it's not going to matter who else is there. If they see within you the quality they think is right, you're going to get the job.

Go to auditions for the experience. It's not a competition, not a test. It's field trip. And somehow you might learn something. You'll bring something back, I guarantee you.

If we were selling vacuum cleaners instead of our talent, one out of ten sales would be considered great. We'd be very successful salesmen. Given the "business" aspect of show business, if we get one part for every ten we go out for, then we are very successful people in our craft.

What thoughts do you have for actors about artistic and psychological challenges in ventriloquism?

All you actors, think about this. If you can make an audience believe that a wooden puppet lives, breathes, expresses, and has life, then what could you do with your human body? How much more depth can you get out of a human body, when you can make an audience believe that a puppet is real? That's the art form. The challenge is to make a character real.

After a performance, people always tell me how remarkable my puppet's reactions were. My puppet's reactions? (Laughs.) My puppet's expression doesn't change. He's carved out of wood. He opens his mouth, moves his head and mouth and eyes in various combinations. He has a wealth of different movements. But he doesn't, for example, wink. He doesn't stick out his tongue. He doesn't make a sour face. He just basically lip syncs.

But an audience will swear that he has smirked, they will swear that he has shown a sour expression. They will swear that he has stuck his tongue out at me while I wasn't looking, and that he's winked at some girl in the audience. He had to wink.

Acting is reaction, that's the key. It's how the ventriloquist reacts to the moment that's going to make the scene work or not. If I react as if my puppet gave me a sour expression, or winked at a girl, the audience will believe it.

Try not to think of ventriloquism as a magic trick. The difference is that a magic trick works on the fact that you don't know how the trick is done. In ventriloquism, you do know how it's done. The trick is no secret. But, you're caught up in the illusion. The difference is, once you find out how the trick works, you cease to be fooled. You appreciate the trick, but you're not fooled by it.

Look at the psychological aspects of ventriloquism. It works on such a high level of psychology that you get caught up in it, even when you know it's not real. It's almost the way we get caught up in a film, knowing it's just a film. We get caught up in it and believe the characters are in jeopardy. Our hearts race when the music starts to rise. We know that's not real, but we suspend our belief.

I remember doing "The Mike Douglas Show" when they set up a microphone for me, and one for my puppet. (Laughs.) Yes, it never occurred to them Bob wouldn't speak. That happens so many times. Even after pointing out Bob's made of wood, they asked if they *should* put a microphone on him. It might look impolite if they didn't at least offer to microphone my co-star. I marvel at the psychological levels this works on.

When a good ventriloquist walks on stage, that suspension of belief happens within minutes. It's fascinating.

What do you think it is about you that allowed you to beat the odds in this business?
In college, when I wanted to be a ventriloquist, everyone would say, "That's kind of like tap dancing, a dead art form." "You're not considering that as a career?" "No one does that today."

Well, I never believed that. Had I known what I know now about the business, maybe I wouldn't have been so starry eyed. But, I always believed there would be a market for what I did. And, my job was really to do it the best I could. To improve doing it, not to be worried about whether I'm going to work or not.

If I listened to all those people, there would not have been a career for me. They would have convinced me that there was just no market for what I did. I believed in myself and in my talent.

This business makes me happy. And there are several different ways I could be happy in the business. It doesn't exclusively have to be acting all the time. It can be producing. It can be writing. Right now, it's my good fortune to be in front of an audience. But I also enjoy producing. I would be happy making a contribution behind the camera, too.

Let me just say these last words to our readers. If you have a talent for something special, work on it. Try to improve it each time it's used. I can't believe there's not a market for your talent. People are bound to appreciate it. As you get better, they'll want to see more. They'll want to know your creative secrets.

Anything done extremely well is a rarity, and rarities are the most valuable commodity in this craft.

Chapter **12** ·

Mel Blanc

Mel Blanc tells us what it was like to work with Jack Benny in radio and television comedy, and traces historical changes in the animated characters he's played. He vividly demonstrates how he created characters like Bugs Bunny, Tweety Bird, and Porky Pig. He also describes the production atmosphere in animation and offers unique insights into the process and techniques of voiceover acting. He performed several of his classic characters for the benchmark film, Who Framed Roger Rabbit? *Mel Blanc was interviewed in his office.*

· · · · · · ·

Career Profile   For over fifty years, Mel Blanc has created the voice of an extraordinary number of America's most renowned cartoon characters. Among them—Bugs Bunny, Porky Pig, Daffy Duck, Sylvester the Cat, Tweety Bird, Yosemite Sam, Road Runner, and Barney Rubble. In appreciation for his contributions to the field, he's received many national honors, including Congressional and Presidential Citations, a star on the Hollywood Walk of Fame, and a permanent Mel Blanc exhibit at the Smithsonian Institution.

Mr. Blanc began his radio career in 1927, as "The Grand Snicker" with the Hoot Owls on Portland, Oregon's KGW. Throughout the Golden Age of Radio (1930s–1940s), he starred in his own show and eighteen other weekly programs. His success in radio put him in league with America's comic radio legends, Jack Benny, Burns and Allen, Judy Canova, and Abbott and Costello.

Warner Brothers Studios had an active animation division in the 1930s, which included the prolific animation team of Friz Freleng, Chuck Jones, and Tex Avery. Mel Blanc was hired to perform the cartoon character voices, at fifteen dollars a week. He received his first screen credit from Warner Brothers, in lieu of a five dollar a week raise.

For fifteen years, Capital Records signed Mr. Blanc to perform his unique brand of comedy. Two of his recordings sold over two million copies ("The Woody Woodpecker Song," "I Tawt I Taw a Puddy Tat"). Total record sales of Mr. Blanc's characters exceeded sixteen million copies.

With the advent of television, Mr. Blanc appeared on "The Jack Benny Show" and did a multitude of voiceover characters for television cartoons and commercials. He and his son, Noel, formed their own production company, Blanc Communications (1960), which is still active in creating commercials and cartoon characters.

By the late 1960s, major studio animation divisions were cut back significantly. Due to economic constraints and changing marketplace trends, the renewed emphasis was on producing television commercials and animated television shows.

Mr. Blanc remained an active force behind a multitude of character voices. In addition to television commercials, he is the singular performer for all Loony Tune character voices and for two animated television series, "Heathcliff" and "The Jetsons."

He voiced many of his classic cartoon characters in the film, *Who Framed Roger Rabbit?* (1988), and recreated the role of Daffy Duck in the full-length animated feature, *Duxorcist* (1988), which celebrated Daffy's fiftieth anniversary.

According to a Harris Research Poll, two characters created by Mr. Blanc are among the nation's favorite—Bugs Bunny and The Road Runner. In a University of Pennsylvania survey, Mr. Blanc was ranked among the top five people that children in America would like to meet. He was in the company of George Washington, Ronald Reagan, and Abe Lincoln.

· · · · · · ·

How did you get started in such a specialized area of acting?
When I was a kid, I loved vaudeville. I went to see Jack Benny and everybody else that came through my hometown, Portland. I hoped that someday I'd grow up and be good enough to be in vaudeville. (Laughs.) But vaudeville died, so I was forced to do nightclubs.

I remember always telling jokes. Even back in grammar school, I had a knack for telling stories. I'd practice a lot of accents and told jokes in different dialects. The kids would get a big laugh out of it. The teachers would laugh too—and then give me lousy grades. (Laughs.)

I started doing professional voiceovers and character voices in the 1930s. About that time, Walt Disney was making the first talking cartoon. Then Warner Brothers came into the act. The studios had a very productive period, churning out enormous numbers of cartoons. I worked for Warners for a long time, doing all the voices for them. I got my first screen credit after negotiating away a five dollar a week raise.

When you first started, was there a lot of competition in this field?
There were several people, but most of them were impersonators. They copied other people's voices. Directors wanted certain voices duplicated, but I wouldn't work that way. To me, copying was like stealing.

Even when a director wanted a certain type of voice, I'd come up with my own version. I'd ask directors to let me try a few lines my own way, and they always liked the end result. That's how I got to do Elmer Fudd. (Elmer's voice) "I twied, and got his voice down pwetty good, just like dat. (Elmer's laugh.)" They loved it. I insisted on doing my own voices.

Historically, there must have been some changes in the characters you've played over the years. Would you discuss some of those changes?
The voices have changed because the characters have actually changed. They're physically different and conceptually different. Porky Pig was initially a real fat little pig. And he just would stutter, so that's all I gave him. But the fact is, the stutter wasn't enough to make him credible. He simply wasn't realistic in the early days.

Leon Schlesinger, who was in charge of Warner Brothers studios at that time, recognized the problem and tried to revive all these "ancient" cartoon characters. If you look at cartoons from the 1920s and 1930s, the characters look like they're a hundred years old. In later versions, they're noticeably younger and better looking. They're also more "childish," which gives them greater identifiability for all ages. You can see the latest streamlining in films like *Who Framed Roger Rabbit?*

You worked with Jack Benny for many years in radio and television. What was his approach to comedy acting?
Jack Benny had a half-hour show, and each Saturday morning he had a read-through. He told his writers what topic he wanted, and they'd bring in a script on that topic. When we did the run-through, we'd read

the parts, and Jack would mark on the script what was funny and what wasn't. Then he'd call the writers and say, "I want all of this changed to this." They listened, made the changes, and somehow, he was always right.

He always had something funny happen to him in the script. Clever, comical things. After he gave his notes to the writers, we'd get a revised script that was full of funny material. Jack only needed eighteen minutes of comedy to choose from, and he always found it. That's how he worked. It was a half-hour show, but he needed eighteen minutes of comedy, so he could turn the rest over for laughs.

And his timing was never wrong.

Jack Benny put you through some acting "challenges." Can you tell us about them?
For nine months, I did the growl for his radio character, Carmichael the Bear. I did the bear, that roar, for *nine months*. Finally I said, "You know Mr. Benny, I can also talk." (Laughs.) Well, he had a good time with that, laughed a lot, then said "Okay, I'll have the writers write something for you."

So they wrote a character who worked in a train station, calling out different railroad tracks. I'd call, "Train leaving on track five for Anaheim, Azuza and Koo... *kamunga!*" Everybody beat me to "Kookamunga." People always thought those were phoney towns, but, they were real.

His writers would always try to stump me, with something I couldn't do. One time, they wrote: "Mel Blanc does an *English* horse whinny." A horse whinny with an English accent?! (Laughs.) I said, "How can you tell the nationality of a horse?" But I never said no. So we finally came to the spot where I was supposed to do the English dialect. I did it like this. (He whinnies with a hint of an English dialect). Well, Jack loved it, "You've done it again. Congratulations, you're now my horse, 'Maxwell.'"

While we were doing his show, he told me "There are two people who are the greatest comics in the business. Jackie Gleason is one. You are the other." Why me? He said, "You couldn't do this show if you weren't able to create so many different roles." I didn't think I was one of the greatest comic actors in the business, but I appreciated him saying that.

You're well known for the characters you created. Can you explain the process you went through to create them? Let's take specific examples—Bugs Bunny, Tweety Bird, Porky Pig, Daffy Duck, Foghorn.

It starts with a picture of the character. The director shows me a picture, and we talk about the character's actions and attitudes.

When they first showed me Bugs Bunny, they told me he was a tough little stinker. They asked for the toughest voice I could get, only small. Make it small. So, I thought, what's the toughest voice in this country? Either Brooklyn or the Bronx. So I put the two of 'em together. I merged the Brooklyn-Bronx accent with a high pitched whine, and that's how I got Bugs Bunny's voice. Personally, he's my favorite character. I'm told he's one of the most listened to characters in the world.

The character of Tweety Bird was created when I tried to find a timid voice for the character of a baby bird. It had to be a tiny voice, but I could speed it up. This sweet little baby bird could be terribly afraid of anything, especially a cat (Tweety's voice): 'I tawt I taw a puddy tat.'' There was another character who spoke like Tweety. He was a little hawk, named Henry. I made him sound like Tweety, but without the baby voice. He spoke straighter, and a little more securely.

Porky Pig was a timid little guy. And he always had trouble talking, so I had him stutter (slips into Porky's stutter): ''tttalking like this... ttthat's all folkss.''

Daffy Duck had trouble talking because of his long beak, so I had him splutter a lot.

I created the character of Foghorn from two people I remembered. One was a character who worked with Fred Allen, the other was a vaudeville comic I saw in Portland. The comic was hard of hearing and exaggerated the problem. He'd yell his dialogue and repeat himself a lot. (Loud character voice) ''Did you hear what I said? Not so loud, I'm not deef!'' I tried to make a nicer fellow of Foghorn, so I added a southern accent. (Character voice) ''I say, pay a— boy, pay attention boy. You looking for chickens? Well you see that little house that says 'D-O-G?' That spells 'chicken.' Go get 'em, boy.''

The most difficult voice to do is Yosemite Sam. Because he's gravelly in the throat. (Character voice) ''My name's Yosemite Sam!'' You can't do too much of that at one time. It hurts the throat.

Do you research accents and dialects for your roles?
I guess I research my roles like any actor would. I listen to people, all the time. I learn dialects and inflections from observing people. Not too many people realize how important inflection is in this business. It's a crucial aspect of creating voices for cartoons.

As an example, for a Japanese accent, I went to an Oriental produce market. There was a Japanese man pointing to a head of lettuce. I said, ''What is that, a cabbage?'' He said, (in dialect) ''Oh, no, awtogether different. That head rettuce." I said, "Oh, the same as this?" And I pointed to the other cabbage. He said, "Oh, no, awtogether different. *That* head of cabbage, *this* head of rettuce." So, I picked up the dialect from him. There's a lot of experimentation in this business, with dialects, voices, inflections.

What's the actual process you go through during production of animated scenes? How much is scripted and how much done ad lib?
First, we go through the story. The director shows me the storyboard, which establishes the character's action in each scene. I can see what the character is going to do, what he's going to say. With that overview, I'll create the different voices.

I've been at this for so many years now that I've gotten to know the characters better than the writers. Sometimes I'd look at the story and tell them, "Bug's wouldn't talk like that." I'd ad-lib something, and they'd say, "Hey, that's much better."

I ad-lib a great deal. If I have a problem with dialogue, I'll change it to find a more appropriate way of saying it. I'll give you an example. Porky Pig is supposed to say, "This sure is a long road." If the line doesn't work, I'll change it, give him something else to look for. (Porky's voice) "This is sure a long roa––, a long roa––, quite a long pavement, huh?"

I switch everything for the character. I do that with practically all the characters. I'll start off following the script, but if a character doesn't work, I'll ad-lib and write my own story, if necessary.

When we're recording, I'm reading the character's lines in the script, but also scanning the next line in the script. That helps me prepare for the next round of dialogue in the scene. I know exactly what type of inflection to use. I give everything I can to my characters: new lines, dialects, inflections.

During production, do you record dialogue sequences as they appear in the script, or do you record each character's voice separately?
Years ago, we did the voices, just as they would appear in the script, one right after the other. If there was a mistake, we'd have to do it all over again. With all the production problems, it took up to *seven hours* to do a half-hour script.

So I suggested we experiment with recording each voice separately. We could do all of Porky Pig's voices, then Daffy Duck's, then Bugs Bunny's. After all the dialogue was recorded, the technician edited it together.

Now, instead of a day and a half to do the voices, it takes one hour or less.

As you create the voice, do you find yourself "becoming" that character?
Yes. As a matter of fact, a friend of mine kept taking pictures of me while I was doing the voice of Bugs Bunny, and Foghorn, and Yosemite Sam. I couldn't understand it. He said, "Mel, everyone of these pictures is of a different voice you do, and the expression in your face is totally different for each one." So, I thought to myself, gee, I must look like Porky Pig sometimes. That's a strange face for a nice Jewish boy. (Laughs.)

When you're in production, how does the director work with you? Is it the same actor-director relationship found in film?
I know a lot of people wonder what a director does in this kind of production. The director is actually directing the voices, the inflections. I'm fortunate, because I can direct myself, and usually get everything down in one take. If I'm not satisfied, I'll ask to do it again.

One of the best directors in this business is Chuck Jones. He's so easy to work with. Chuck tells me what he wants, suggests how the character talks. Then I try it in the character's voice. the relationship is very positive.

I like directors who know what they want and work with me to get it. Tex Avery is a director who's quite marvelous with voices. Friz Freleng, who ran Warner Brothers Cartoons for many years, is a terrific director, but he doesn't know how to direct *me*. When I'd start a character's voice, Friz would immediately stop me, "That's not right." Finally, I'd convince him to try it my way. As soon as I created the voice, he'd reconsider, "That's it!" After all that tumult, we end up doing my voice, anyway. Not only that, but we complete it on the first take! But he *is* a good director.

Is there any character inside Mel Blanc that hasn't had a chance to come out yet?
Thousands of them! And I've got to keep them under control. When I speak at colleges, I'm always asked to try new voices. One fellow challenged me to do a rhinoceros. He said, "You can't do a rhinoceros, can ya?

I said, "Hmm. Let's see. A rhinoceros. He's got to be big. And heavy (Character voice comes in), and fat. He's in the water, and he sort of spits when he speaks.'' The students were amazed, ''Hey, that really sounds like a rhinoceros.''

They love to hear me experiment with new stuff, but it's always the old standards I'm called on to do.

Did you ever regret not having performed more roles in front of the camera?
My voice was everywhere—radio, television, film, records—but my face was anonymous. That's what I regretted. People didn't know me. My characters were internationally known, but I'd walk down the street and people didn't know who the hell I was.

For that reason, I enjoyed doing an American Express commercial, where I played myself. The commercial included my voices, my characters, and played up my physical ''anonymity.'' After the commercials aired, I'd walk down the street, and *everyone* recognized me.

Oh, the amazing power of sixty-second commercials!

What advice would you give to someone who wants to do what you do? What specific acting skills or vocal training would you recommend?
First, practice building your imagination. Be able to create and see imaginary characters. That's what I've told students at hundreds of colleges.

A simple exercise is to imagine an animal, and slowly give it life, find a voice, an attitude. Pick any animal for practice. Maybe a little kitten. What does a kitten do? Observe it, watch it, create it, picture it. Then, go with it. The little kitten ''meows.'' It's tiny, so it has a real small voice.

You can even give the character a foreign accent for a challenge. What would this little kitten sound like if it could talk Italian? French? English? We know it's real small, so the voice has to be very high. And we can make it sound like (meows in character voice), ''It's a foreign pussy cat.''

As for vocal skills, the trick is to broaden your natural range. I urge people to stretch their vocal range. Practice every day. Exercise. Strengthen your voice, larynx, esophagus. Try anything to stretch your vocal range. Do a real high voice, then a deep, low voice.

The more you do, the more natural it will be.

What do you think it is about Mel Blanc that enabled you to beat the odds in this business?

First of all, I'm pretty. (Laughs.) Actually, I've got a friendly face, and a friendly voice. I worked hard to keep my voice friendly. I know people enjoy hearing the characters I create.

As for the vocal qualities, God was good to me. I have a "rubber voice." (Laughs.) That's as good an explanation as anything. I can go as high as I want (speaks in falsetto voice); or as low as I want (speaks in deep voice). I don't know if the magic lies in my esophagus or larynx. But I'm lucky to have such a naturally wide range.

Now I think it's only appropriate for us to close this chapter with some words of advice to your readers from Porky Pig. (Porky's voice) "You can do it if ya try. . ! Just give it ya besst . . . And ttthat's all fffolkss . . !"

Selected and

Annotated Bibliography

This bibliography includes resources cited in the text. In addition, you'll find works that cover relevant topics of interest, from the history of acting to the techniques of television, film, and stage performance.

. . . .

Adler, Stella. "The Reality of Doing," *Tulane Drama Review,* 9, No. 2 (Fall, 1964): 136–155. Adler is interviewed, with Vera Soloviova and Sanford Meisner, about realistic acting techniques in America.

Barr, Tony. *Acting for the Camera.* New York: Harper & Row, 1982. This explains the realistic acting techniques used by the Film Actors Workshop in Los Angeles. The author is Vice President of Drama at CBS-TV and is founder of the Film Actors Workshop.

Blum, Richard A. *American Film Acting: The Stanislavski Heritage.* Ann Arbor: U.M.I. Research Press, 1984. This book provides an historical overview of the Stanislavski system, tracing the theories and controversies from New York to Hollywood.

Boleslavski, Richard. *Acting: The First Six Lessons.* New York: Theatre Arts Books, 1949. The lessons in this book were based on early Stanislavski principles: concentration, emotional memory, dramatic action, characterization, observation, rhythm.

Clurman, Harold. *The Fervent Years: The Story of the Group Theatre and the Thirties.* New York: Hill and Wang, 1945. This is the most complete history of the Group Theatre and its members, by one of its foremost members.

Cohen, Robert. *Acting Professionally: Raw Facts About Careers in Acting.* Palo Alto, California: Mayfield Publishing, 1975. A straightforward, practical guide to the relevant facts about careers in stage and film acting.

Cole, Toby and Chinoy, Helen K., eds. *Actors on Acting.* New York: Crown Publishers, Rev., 1970. A very important resource for understanding world acting techniques and styles. This is a compilation of many views by different actors of different times.

Diderot, Denis. "The Paradox of Acting," in William Archer, *Masks or Faces.* New York: Hill and Wang, 1957. Diderot (1830) believed actors can portray emotions without feeling them.

Dmytryk, Edward. *On Screen Acting.* Boston: Focal Press, 1986. Written with actress (and wife) Jean Porter, this is a lively exchange of views about the work of film actors.

Duerr, Edwin. *Length and Depth of Acting.* New York: Holt, Rinehart and Winston, 1962. Offers scholarly insights into acting styles throughout the world.

Easty, Edward. *On Method Acting.* New York: House of Collectibles, 1966. This small volume clearly and concisely explains the techniques and exercises used at the Actor's Studio. The book is used by actors in training at the Lee Strasberg Theatre Institute.

Everson, William K. *American Silent Film.* New York: Oxford University Press, 1978. This offers an historical perspective of the silent film era and early film acting.

Garfield, David. *A Player's Place: The Story of the Actors Studio.* New York: Macmillan Publishing, 1980. This book is a well documented and informative history of the Actor's Studio. The author is a member of the Studio.

Gillette, William. *The Illusion of the First Time in Acting. Papers on Acting I.* New York: Dramatic Museum of Columbia University, 1915; rpt. 1958. This is the first argument for the concept of creating an illusion of reality in acting.

Hagen, Uta. *Respect for Acting.* New York: Macmillan Co., 1973. This is a well-received book that inspires as well as explains a great deal about the acting craft. The author is an actress and teacher in New York.

Kazan, Elia. *Elia Kazan: A Life.* New York: Knopf, 1988. A vast volume that encompasses years of professional experience. As a director, Kazan talks about problems and conflicts in the Group Theatre, the Actor's Studio, and with contemporary film actors.

Lewis, Robert. *Method or Madness?* New York: Samuel French, 1958. A nuts and bolts discussion of the Method, with delineation of strengths and weaknesses.

————. *Advice to the Players.* New York: Harper & Row, 1980. An insightful and helpful book on acting, dealing with both theory and practice.

Moore, Sonia. *The Stanislavski System: The Professional Training of an Actor.* New York: The Viking Press, 1974. A very accessible and informative small volume that clearly explains the basics of the Stanislavski system.

Naremore, James. *Acting in the Cinema.* University of California Press, 1988. A survey and history of performing arts styles in film.

Olivier, Laurence. *On Acting.* New York: Simon and Schuster, 1986. Olivier is one of this generation's greatest actors. This books is particularly illuminating about his personal acting style in theatre, film, and television.

Probst, Leonard, ed. *Off-Camera: Leveling About Themselves.* New York: Stein and Day, 1975. Informal interviews with major film stars, including Al Pacino, Paul Newman, George C. Scott, Dustin Hoffman, Lynn Redgrave, Shirley MacLaine, Angela Lansbury, and Woody Allen.

Pudovkin, V.I. *Film Technique and Film Acting: The Cinema Writings of Pudovkin.* Translated by Ivor Montagu. London: Vision Press, 1954. Pudovkin thought film production techniques demanded a new intimacy and naturalism in acting.

Stanislavski, Constantin. *An Actor Prepares.* Translated by Elizabeth Reynolds Hapgood. New York: Theatre Arts Books, 1936. This work outlines the basic principles of finding inner truth, character objectives, super-objectives, and throughline of action.

———. *Building a Character.* Translated by Elizabeth Reynolds Hapgood. New York: Theatre Arts Books, 1949. This volume concentrates on "outer technique." It was not as widely known, initially, as the other works.

———. *Creating a Role.* Translated by Elizabeth Reynolds Hapgood. New York: Theatre Arts Books, 1961. This covers script analysis and character development techniques.

———. *My Life in Art,* 2nd ed. Translated by J.J. Robbins. New York: Theatre Arts Books, 1924. This is Stanislavski's autobiography, including reflections and observations about acting.

———. *Stanislavski's Legacy: A Collection of Comments on a Variety of Aspects of an Actor's Art and Life.* Edited and translated by Elizabeth Reynolds Hapgood. New York: Theatre Arts Books, 1958. A compilation of important ideas and insights about the art of acting.

Strasberg, Lee. *A Dream of Passion: The Development of the Method.* New York: Little, Brown & Company, 1987. A posthumously published account of Strasberg's evolution of the Method. Includes Strasberg's perspectives on the Actor's Studio, the Method, and acting styles in America.

Talma, Francois Joseph. "Reflections on Acting," in *Papers on Acting,* edited by Brander Matthews. New York: Hill and Wang, 1958. Talma (1877) thought actors must feel emotions to make them seem more realistic.

Index